W9-CHL-384

Makoto Komiyama, Toshifumi Takeuchi,
Takashi Mukawa, Hiroyuki Asanuma
Molecular Imprinting

Makoto Komiyama,
Toshifumi Takeuchi,
Takashi Mukawa,
Hiroyuki Asanuma

Molecular Imprinting

From Fundamentals to Applications

Prof. Makoto Komiyama
Research Center for Advanced Science and Technology
The University of Tokyo
4-6-1 Komaba, Meguro-ku,
Tokyo 153-8904
Japan

Prof. Toshifumi Takeuchi
Graduate School of Science and Technology
Kobe University
1-1 Rokkodai-cho, Nada-ku,
Kobe 657-8501
Japan

Prof. Takashi Mukawa
Graduate School of Science and Technology
Kobe University
1-1 Rokkodai-cho, Nada-ku,
Kobe 657-8501
Japan

Prof. Hiroyuki Asanuma
Research Center for Advanced Science and Technology
The University of Tokyo
4-6-1 Komaba, Meguro-ku,
Tokyo 153-8904
Japan

Library of Congress Card No. applied for.

British Library Cataloguing-in-Publication Data: A catalogue record for this book is available from the British Library

Deutsche Bibliothek Cataloguing-in-Publication Data: A catalogue record for this publication is available from the Die Deutschen Bibliothek

Printed in the Federal Republic of Germany

Printed on acid-free paper.

Composition TypoDesign Hecker GmbH, Leimen
Printing Strauss Offsetdruck GmbH, Mörlenbach
Bookbinding Litges & Dopf GmbH, Heppenheim
Cover Design G. Schulz, Fussgönheim

ISBN 3-527-30569-6

Contents

Preface

"Molecular imprinting" is a newly developed methodology which provides molecular assemblies of desired structures and properties. In the presence of a template molecule, functional monomers are polymerized and immobilized complementarily to this molecule. After the polymerization, the template is removed. During these procedures, a snapshot of the system is taken so that the resultant molecular assembly exclusively binds this template as well as its analogues. This is a kind of challenge towards the second law of thermodynamics saying that "the entropy of the universe should necessarily increase". No one doubts that this method is one of the keys for future science and technology.

The term "imprinting" is reminiscent of psychological phenomena in nature. A newly-hatched chick of some birds (e.g., wild duck) gets an overwhelmingly strong impact from the object that it encounters first after its birth. Thus, it believes that this newcomer should be its own parent. Anything moving can be this "*a posteriori* parent". Other birds, animals, and even human beings are acceptable. Our "molecular imprinting", which occurs between molecules, is also versatile in scope.

This book is written primarily as a textbook for graduate courses. Accordingly, fundamentals of "molecular imprinting" are described in detail. Even undergraduate students should be able to understand the whole context and have a clear idea on this elegant methodology. Experimental details are presented in many reaction examples so that the readers can repeat these experiments and also use this method for their own research. Furthermore, most important recent progresses are covered in chapters 7 and 8. These parts should be very informative even for advanced-course students and researchers who can overview this rapidly growing area and get valuable hints for their future work.

We should like to thank a number of people who have worked for the development of the "molecular imprinting method". Our sincere appreciation is also extended to the staff members and students in our laboratories for helping with tables and figures, and to our wives, Mitsuko Komiyama, Emiko Takeuchi, and Maki Asanuma, whose understanding was essential. However, as usual, the final criticism must be borne by us.

May 2002

Makoto Komiyama
Research Center for Advanced Science and Technology University of Tokyo, Tokyo, Japan

Chapter 1
Introduction

1.1
Importance of Receptor Molecules in Advanced Science and Technology

In solutions and gases (but not solids), we know that most of the molecules are randomly moving around. Each molecule does not much care about its neighbors and behaves as it wishes. Complexes between the molecules are formed only through accidental encounters between them. The lifetimes of these collision complexes are negligibly small, and their concentrations in solutions (or in gases) are virtually nil. However, some kinds of molecules (»receptor molecules« or simply »receptors«) precisely differentiate between one molecule and another. They exclusively pick out their own partner molecule from a number of molecules in the system and form a non-covalent complex with this molecule. These complexes are sufficiently stable, and their equilibrium concentrations are considerable. All the molecules other than the partner are completely neglected here, just as we easily and precisely find our friend even in a crowd of people at the station entrance and go to dinner with him or her. When necessary, predetermined reactions and/or catalyses take place in these non-covalent complexes, as observed in enzymatic reactions. Such discrimination between molecules is called »molecular recognition«, and is one of the essential keys to the existence of living things [1].

In the latest science and technology, the importance of »receptors« and »molecular recognition« has been growing rapidly. This is mainly because one molecule is now a functional unit and has its own role. Highly complicated operations are achieved by a combination of the functions of each molecule. In order to develop highly sophisticated systems under these conditions, we should align a number of molecules in a predetermined manner and allow each to perform its own function. Here, of course, all the molecules must know who are their neighbors, what physicochemical properties they have, and what these neighbors are doing at any moment. This situation is entirely different from ones where bulk properties of materials, rather than the functions of each molecule therein, were the main concerns.

A »molecular imprinting method« has recently been developed to provide versatile receptors efficiently and economically. In principle, movements of molecules are frozen in polymeric structures so that they are immobilized in a desired fashion. This method is so unique and challenging that the scope of future applications is hard to predict precisely at the present time. This chapter deals with the current status of relevant sciences so that the reader can confirm the importance of the molecular imprinting method.

1.2
Naturally Occurring Receptors

There are many cells and molecules in our body, and all of them are cooperatively working in an enormously ordered fashion. Without such mutual understanding and cooperation, we cannot survive. Thus, molecular recognition is essential for the existence of life. For example, the receptors on the surface of cell membranes bind hormones and are responsible for cell-to-cell communication. When the receptor binds a hormone, its conformation is changed and the message of the hormone (e.g., lack of glucose in the body) is passed to the cell in terms of this conformational change. Now that this cell knows what is required in the body at that moment, it promotes (or suppresses) the corresponding bioreaction(s) to respond to this requirement appropriately. In the above example, glycogen is hydrolyzed and glucose is supplied to the body. The most important thing in these systems is that one receptor accepts only one specific hormone and never significantly interacts with others. Furthermore, this receptor/hormone interaction is enormously strong. Thus, even small amount of hormone can correctly deliver its information to the target cell without information cross talk between cells. On the other hand, selective guest binding by antibodies is essential for our immune response. These proteins patrol around in our body like policemen, arrest a foreign substance (antigen) when it enters the body, and take it to a lysosome (a cell organelle) where the antigen is destroyed. Our body is successfully protected. As would be expected, the differentiation by an antibody between the target antigen and the others (and also between foreign substances and the intrinsic ones in our body) must be rigorously strict.

As is well known, enzymes also show high substrate specificity. Each of them exclusively chooses a certain substrate (specific substrate) and transforms it into a predetermined product. All other compounds in the system (even if they resemble the specific substrate) are kept intact. Another enzyme takes another specific substrate and executes a different mission. This substrate specificity primarily comes from selective guest binding by substrate-binding sites of enzymes. Furthermore, only the specific substrate is efficiently transformed into the desired products, since the catalytically active amino acid residues of enzymes, located near the substrate-binding sites, are arranged suitably only for this transformation.

Detailed information on molecular recognition in nature is now available from X-ray crystallography and NMR spectroscopy [3]. The substrate-binding sites of enzymes are apolar pockets or clefts, which are formed from a number of amino acid residues. There, several functional groups (OH, NH_2, COOH, imidazole, main-chain amide groups, and others) are precisely placed to interact with the functional groups of a specific substrate. For example, an ammonium ion of an enzyme shows Coulomb interaction with a negatively charged carboxylate of its specific substrate. A hydrogen bond is formed between the OH residues of the enzyme and the substrate. Furthermore, apolar binding occurs between them. For the specific substrate only, all these interactions satisfactorily and cooperatively operate, and a stable non-covalent complex is formed. Antibody-antigen interactions, as well as guest binding by membrane-receptors, occur essentially in the same manner. A number of amino acid residues of antibodies (or receptors) are oriented complementarily to the functional groups of the target antigens (or hormones). Precise molecular recognition also occurs when proteins interact with each other.

1.3
Artificial Receptors

The elegance of molecular recognition in nature has been spurring many scientists to mimic it. One of the greatest advantages of artificial receptors over naturally occurring ones is freedom of molecular design. Their frameworks are never restricted to proteins, and a variety of skeletons (e.g., carbon chains and fused aromatic rings) can be used. Thus, the stability, flexibility, and other properties are freely modulated according to need. Even functional groups that are not found in nature can be employed in these man-made compounds. Furthermore, when necessary, the activity to response towards outer stimuli (photo-irradiation, pH change, electric field, and others) can be provided by using appropriate functional groups. The spectrum of functions is far wider than that of naturally occurring ones.

Pioneering works by Cram, Lehn, and Pedersen (Nobel Prize winners in 1987) established that the following factors are necessary for accurate molecular recognition [1].

1. Functional residues of guest and receptor must be complementary to each other.
2. Conformational freedom of both components should be minimized.
3. Chemical circumstances should be appropriately regulated.

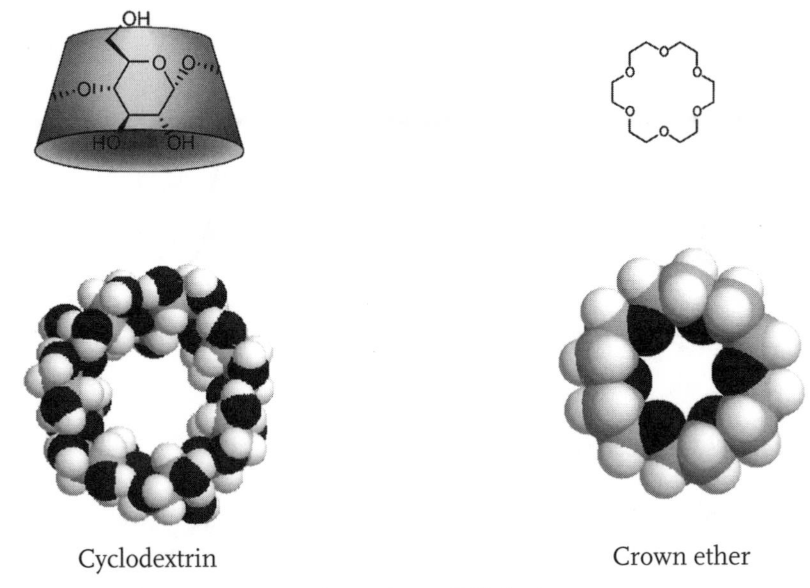

Fig. 1-1 Typical cyclic host molecules used as scaffolds for the recognition of specific target guests

In many cases, various functional residues are covalently attached to cyclic host molecules (e.g., cyclodextrins and crown ethers; see Fig. 1.1). Although each of the interactions (hydrogen-bonding, electrostatic, and apolar-binding) is rather weak, remarkably high selectivities and binding strengths are accomplished when all of them work cooperatively. Alternatively, functional residues are bound to carbon skeletons so that they converge to the central guest binding portion. A typical example of this type of host molecules is presented in Fig. 1.2. In principle, good receptors can be successfully synthesized, as long as (i) we are not concerned about cost and time for their preparation, (ii) our target guest is fairly small, and (iii) we can use organic solvents for the guest recognition. However, these conditions are hardly ever fulfilled, making the molecular imprinting method significant and attractive.

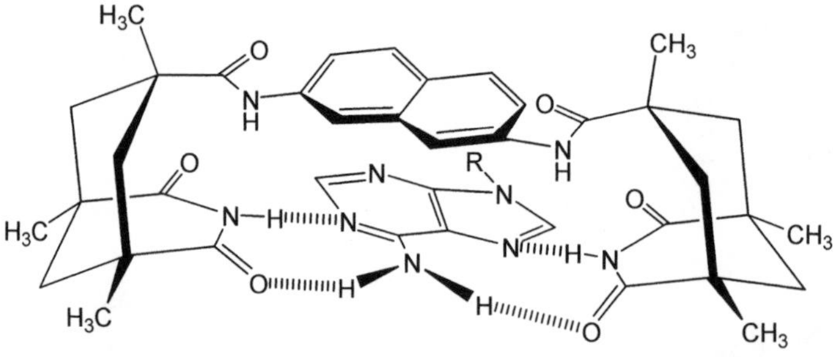

Fig. 1-2 Convergent host molecule for the recognition of adenine derivative [4]

1.4
Receptors for Practical Applications

In industry, receptors are being used to separate the target product economically from reaction mixtures and to remove dangerous chemicals from waste-water. Applications to molecular biology (regulation of bioreactions, separation of biomaterials, and others) are also promising. In some cases, the cost of the separation of the product and its purification accounts for more than half of the total cost of production. Thus, highly selective and economical receptors are crucially important for successful business. Furthermore, unprecedentedly compact and sophisticated devices are fabricated by placing molecules in an ordered fashion. Here, precise recognition between molecules is essential. Molecular memories, molecular devices, and molecular machines have been already realized to some extent [2]. It is also said that artificial cells could be prepared in the near future by placing man-made receptors on artificial cell membranes.

When we design new receptors for future applications, the following factors must be carefully considered [5]:

1. Easy preparation in large amounts at low cost.
2. Stability and activity under wide operation conditions.
3. Selective and strong binding toward large guest molecules.
4. Guest binding in water.

The first and the second requirements are trivial. The third one comes from the fact that important molecules in advanced sciences (proteins, nucleic acids, polysaccharides, bioactive chemicals, and others) are usually large. The fourth requirement has been rapidly increasing in importance, since economical, ecological, and environmental restraints are spurring the replacement of organic solvents with water. For biotechnology, of course, the solvent must be water.

1.5 Why is the Molecular Imprinting Method so Promising?

The molecular imprinting method is quite simple and easy to perform in a tailor-made fashion. All we need are functional monomers, templates, solvents, and crosslinking agents. Polymerization is followed by the removal of the template. During these processes, a number of functional monomers are assembled in an orderly fashion and their functional groups are placed at the desired sites in the cavities of desired size. Neither complicated organic synthesis nor complex molecular design is necessary. If you wish to have a receptor toward a certain guest compound, you can simply polymerize appropriate monomer(s) in the presence of this guest compound (or its analog) as the template. For another guest, you can use another combination of functional monomer and template. The corresponding receptor can be at hand almost automatically. Requirements 1 and 2 would seem to be sufficiently fulfilled. Furthermore, significant progress has recently been made in the fulfillment of requirements 3 and 4 (see Chapter 8). This methodology certainly opens the way to further developments in science and technology.

References

1 J.-M. Lehn, *Supramolecular Chemistry*, VCH, Weinheim, 1995.

2 (a) D. B. Amabilino, J. F. Stoddart, *Chem. Rev.* **1995**, *95*, 2725. (b) T. R. Kelly, H. Silva, R. A. Silva, *Nature* **1999**, *401*, 150. (c) N. Koumura, R. W. J. Zijlstra, R. A. Delden, N. Harada, B. L. Feringa, *Nature* **1999**, *401*, 152. (d) H. Shigekawa, K. Miyake, J. Sumaoka, A. Harada, M. Komiyama, *J. Am. Chem. Soc.* **2000**, *122*, 5411.

3 L. Stryer, *Biochemistry*, 3rd edn, W. H. Freeman and Co., New York, 1988.

4 Rebek, J. Jr et al., *J. Am. Chem. Soc.* **1987**, *109*, 5033.

5 H. Asanuma, T. Hishiya, M. Komiyama, *Adv. Mater.* **2000**, *12*, 1019.

Chapter 2
Fundamentals of molecular imprinting

2.1
Introduction

As described in Chapter 1 (Section 1.3), a number of elegant receptor molecules have already been synthesized by aligning functional groups on appropriate scaffolds. Typical examples are presented in Figs. 1.1 and 1.2. They have well-defined molecular structures, and satisfactorily show both high selectivity and high binding activity toward the target guest compound. Detailed and fundamental knowledge of molecular design of sophisticated receptors has also been accumulated through these studies. These trends should go on further and provide still more fruitful results. From the viewpoints of practical applications to industry and our daily lives, however, these synthetic receptors have several drawbacks. The first is rather poor availability. Their synthesis often requires five or more reaction steps, and only a small amount (e.g., 1 g) is eventually obtained. Thus, they are usually too expensive for common industrial uses. Secondly, the design of receptors for large guest molecules is quite difficult, since the scaffolds available are in most cases smaller than several angstroms. Under these conditions, it is hard to place two or more functional groups which are located at notably remote sites (e.g., > 10 Å), even for highly skilled organic chemists using the most advanced organic chemistry. Thirdly, it is difficult to provide these synthetic receptors with an appropriate reaction field for precise molecular recognition (without changing the media used). Precise molecular recognition in water is especially difficult, since hydrogen bonds, used in the recognition by naturally occurring receptors, are easily broken because of competition with the water. As is observed in the natural receptors, polymeric structures are usually necessary to form hydrogen bonds in bulk water.

The molecular imprinting method is the most promising solution to these problems. Simply by polymerizing appropriate functional monomers in the presence of a template, desired receptors are cheaply prepared in tailor-made fashion and on a kilogram scale (even on a ton scale). Receptors for large templates are also easily obtainable. No complicated organic synthesis is necessary. Furthermore, the chemical circumstances in guest-binding sites are easily regulated by combining appropriate monomers, crosslinking agents, and/or comonomers. These features make molecular imprinting method one of the most attractive methodologies.

2.2
General Principle of Molecular Imprinting

Suppose that a number of functional molecules interact with a template molecule in solution (or in a gas). The interactions are hydrogen-bonding, electrostatic, apolar, and any other non-covalent interactions. Here, these functional molecules are arranged in an orderly manner with respect to each other so that their functional groups are complementary to the template. Then, what happens if we suddenly remove the template molecule from this system? As you can easily imagine, all the functional molecules would soon start moving randomly, and their ordering would disappear. As a result, the memory about the template is immediately gone. In the molecular imprinting method, however, this randomization is minimized by connecting these functional molecules together by means of a polymer backbone (Fig. 2.1). A kind of snapshot of solution (or gas) is taken, and the structure of the template is memorized in these polymers, providing the target receptors.

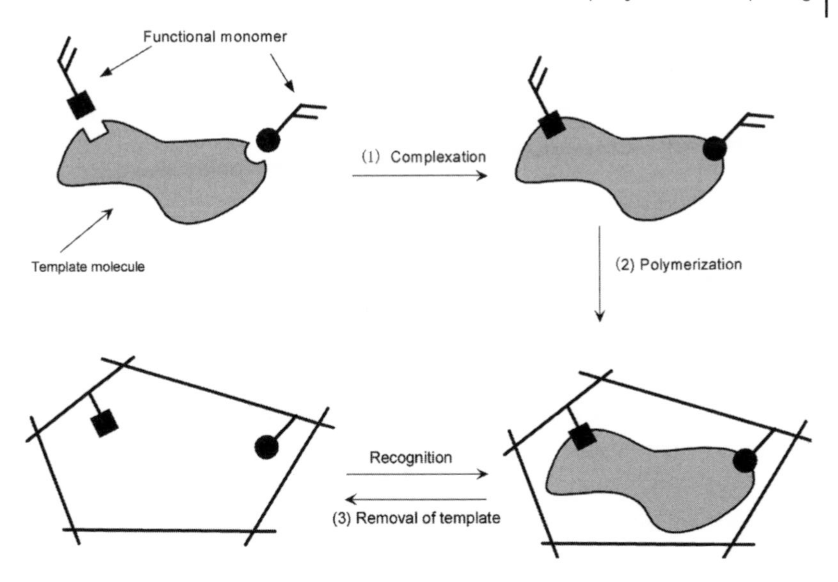

Fig. 2-1 Schematic illustration of molecular imprinting

Molecular imprinting processes are composed of the following three steps:

1. Preparation of covalent conjugate or non-covalent adduct between a functional monomer and a template molecule,
2. Polymerization of this monomer-template conjugate (or adduct), and
3. Removal of the template from the polymer.

In step 1, functional monomer and template are connected by a covalent linkage (in »covalent imprinting«) or they are placed nearby through non-covalent interactions (in »non-covalent imprinting«). In step 2, the structures of these conjugates (or adducts) are frozen in a three-dimensional network of polymers. The functional residues (derived from the functional monomers) are topographically complementary to the template. In step 3, the template molecules are removed from the polymer. Here, the space in the polymer originally occupied by the template molecule is left as a cavity. Under appropriate conditions, these cavities satisfactorily remember the size, structure, and other physicochemical properties of the template, and bind this molecule (or its analog) efficiently and selectively.

2.3 Covalent Imprinting and Non-covalent Imprinting

As described above, the molecular imprinting method is of two types, depending on the nature of adducts between functional monomer and template (either covalent or non-covalent). Typical examples of these two kinds of methods are presented in Figs. 2.2 and 2.3. Both have advantages and disadvantages, and thus the choice of the best method strongly depends on various factors (see below).

1. Covalent Imprinting (Fig. 2.2)

Prior to polymerization, functional monomer and template are bound to each other by covalent linkage (step 1). Then, this covalent conjugate is polymerized under the conditions where the covalent linkage is intact (step 2). After the polymerization, the covalent linkage is cleaved and the template is removed from the polymer (step 3). Upon the guest binding by the imprinted polymers, the same covalent linkage is formed (note that an improved strategy is presented below in Fig. 2.4).

2. Non-covalent Imprinting (Fig. 2.3)

In order to connect a functional monomer with a template, non-covalent interactions (e.g., hydrogen bonding, electrostatic interaction, and coordination-bond formation) are used here. Thus, the adducts can be obtained *in situ* simply by adding the components to reaction mixtures (step 1). After the polymerization (step 2), the template is removed by extracting the polymer with appropriate solvents (step 3). The guest binding by the polymer occurs through the corresponding non-covalent interactions.

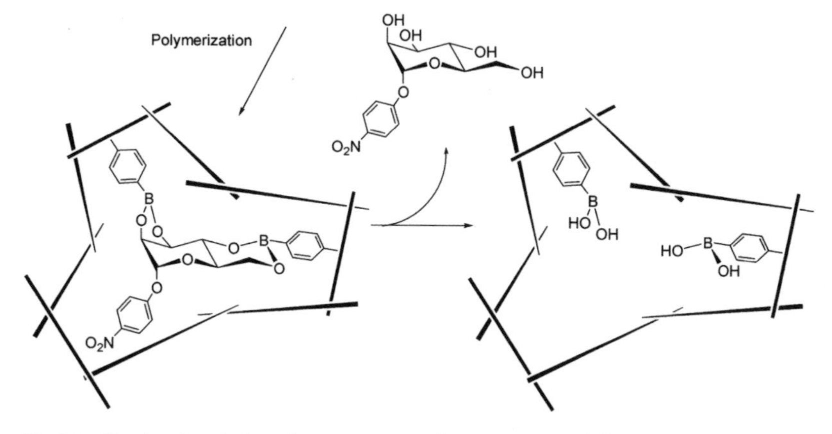

Fig. 2-2 Covalent imprinting of mannopyranoside using its 4-vinylphenylboronic acid ester as a functional monomer

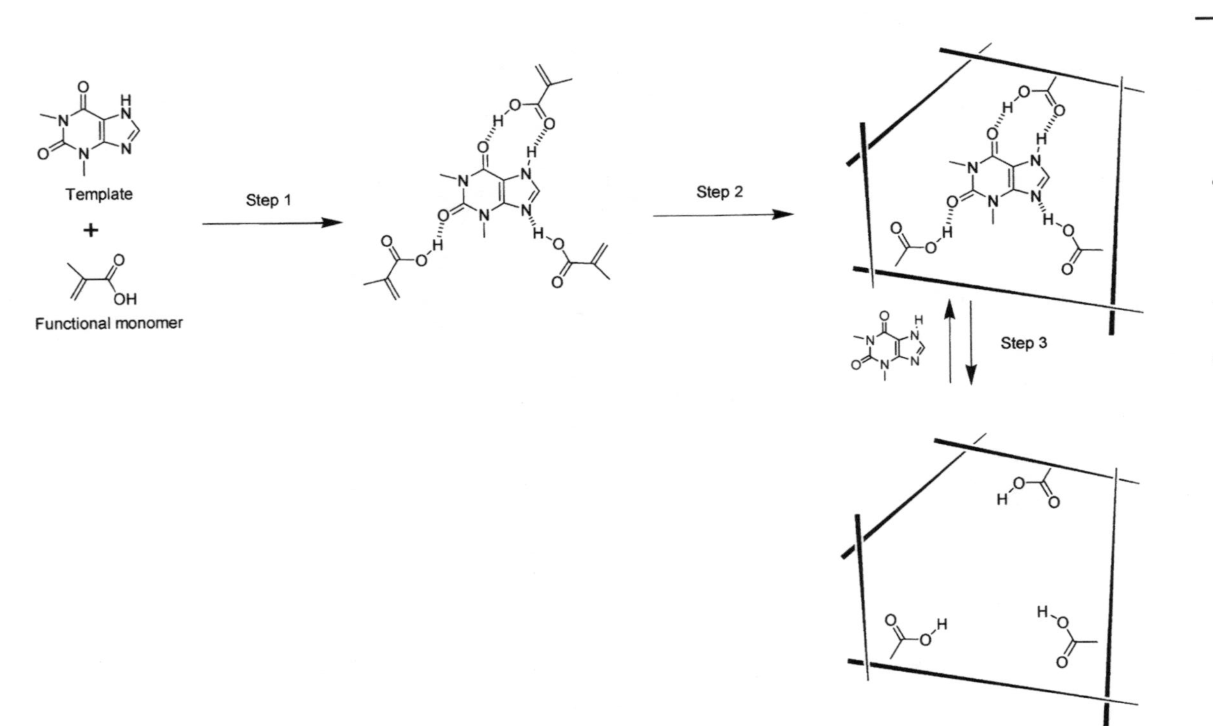

Fig. 2-3 Non-covalent imprinting by theophylline (a drug):
Step1: Pre-organization of functional monomers through non-covalent interactions
Step2: Polymerization of pre-organized functional monomers.
Step 3: Removal of the template

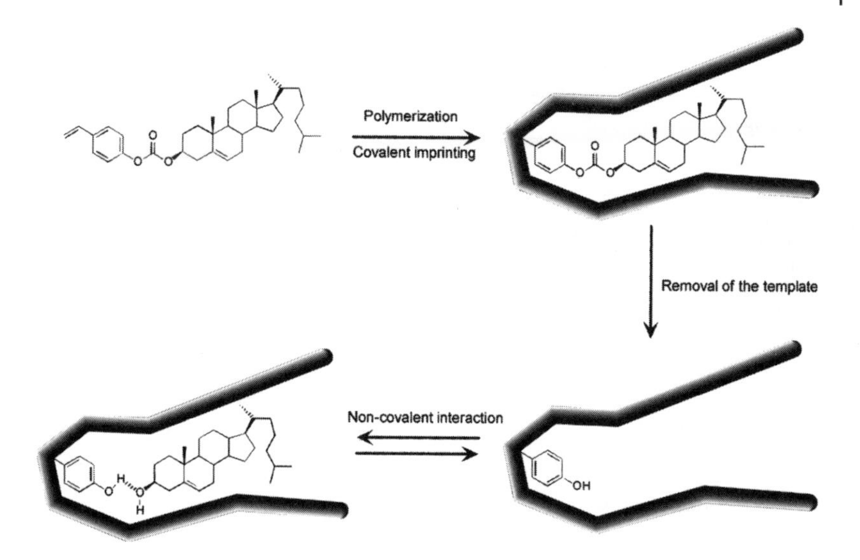

Fig. 2-4 Hybridization of covalent imprinting and non-covalent imprinting

2.4
Advantages and Disadvantages of Covalent and Non-covalent Imprinting

In general, non-covalent imprinting is easier to achieve and applicable to a wider spectrum of templates. With respect to the strictness of imprinting, however, covalent imprinting is usually superior. Other characteristics are described in Table 2.1. One should choose either of these two methods, depending on the need and situation of their operations (kind of the target guest compound, the guest selectivity required, the cost and time allowable for the preparation, and others).

Table 2-1 Advantages and disadvantages of covalent and non-covalent imprinting

	Covalent	***Non-covalent***
Synthesis of monomer-template conjugate	necessary	unnecessary
Polymerization conditions	rather free	restricted
Removal of template after polymerization	difficult	easy
Guest-binding and guest-release	slow	fast
Structure of guest-binding site	clearer	less clear

1. Covalent Imprinting

Advantages:

1. Monomer-template conjugates are stable and stoichiometric, and thus the molecular imprinting processes (as well as the structure of guest-binding sites in the polymer) are relatively clear-cut.
2. A wide variety of polymerization conditions (e.g., high temperature, high or low pH, and highly polar solvent) can be employed, since the conjugates are formed by covalent linkages and are sufficiently stable.

Disadvantages:

1. Synthesis of the monomer-template conjugate is often troublesome and less economical.
2. The number of reversible covalent linkages available is limited.
3. The imprinting effect is in some case diminished in step 3 (cleavage of covalent linkages), which requires rather severe conditions.
4. Guest binding and guest release are slow, since they involve the formation and breakdown of a covalent linkage.

2. Non-covalent Imprinting

Advantages:

1. Synthesis of covalent monomer-template conjugates is unnecessary.
2. Template is easily removed from the polymer under very mild conditions, since it is only weakly bound by non-covalent interactions.
3. Guest binding and guest release, which take advantage of non-covalent interactions, are fast.

Disadvantages:

1. The imprinting process is less clear-cut (monomer-template adduct is labile and not strictly stoichiometric).
2. The polymerization conditions must be carefully chosen to maximize the formation of non-covalent adduct in the mixtures.
3. The functional monomers existing in large excess (in order to displace the equilibrium for adduct-formation) often provide non-specific binding sites, diminishing the binding selectivity.

2.5 History

In the past, chemists were familiar with so-called »template effects«, although the term »molecular imprinting« was not used. When some molecules are added to reaction mixtures, chemical reactions are accelerated and/or product distribution is notably altered. Here, these additives interact with the reactants, and place them in mutual conformations which efficiently lead to one of the possible products. Alternatively, they stabilize certain products and promote their formation. Apparently, these phenomena are associated with the concept of molecular imprinting. However, it was the pioneering works by Wulff, Shea, Mosbach, and others that made the »molecular imprinting method« as popular as it is today.

1. Covalent Imprinting

Wulff and his coworkers reported the first covalent imprinting in 1977 (Fig. 2.2) [1]. They synthesized 2:1 covalent conjugate of *p*-vinylbenzeneboronic acid with 4-nitrophenyl-α-D-mannopyranoside (the template), and copolymerized this conjugate with methyl methacrylate and ethylene dimethacrylate (a crosslinking monomer). After the polymerization, the boronic acid ester in the polymer was cleaved, and the 4-nitrophenyl-α-D-mannopyranoside was removed. Exactly as desired, the resultant polymer strongly and selectively bound this sugar. The mutual conformation of the two boronic acid groups in the covalent conjugate were frozen in the polymer, and the structure of the template was memorized. Similarly, Shea formed a ketal conjugate between the carbonyl group of a template and the 1,3-diol group in a functional monomer, and used this covalent conjugate for molecular imprinting [2].

2. Non-covalent Imprinting

Mosbach and his coworkers showed that covalent linkages between functional monomer and template are not necessarily required for molecular imprinting, and even non-covalent interactions between them work sufficiently [3, 4]. Simply by mixing them in reaction mixtures, their non-covalent adducts were spontaneously formed, and satisfactory imprinting effects were obtained. In the imprinting of methacrylic acid with theophylline (a drug), for example, a non-covalent monomer-template adduct was formed through hydrogen bonding and electrostatic interaction (Fig. 2.3). The same strategy was successful for the imprinting with various drugs, insecticides, and other practically important chemicals. Many experimental workers were surprised to see that the methods are so simple while the imprinting effects are so remarkable. They were soon persuaded that this method is satisfactorily applicable to a wide range of molecules and started to use it in their own laboratories.

3. Hybridization of Covalent Imprinting and Non-covalent Imprinting
The advantage of covalent imprinting (clear-cut nature) and that of non-covalent imprinting (fast guest binding) were combined [5]. The polymers were prepared as in covalent imprinting, but the guest binding employed non-covalent interactions (see Fig. 2.4). One of the shortcomings of covalent imprinting (slow guest binding and guest release) has been solved by this approach. Of course, many other improvements have also been attempted as described in this book.

Nowadays, the molecular imprinting method is being widely used to place the target molecules at desired sites easily and economically. The purpose is not restricted to the preparation of artificial receptors and is much wider in scope. There are no alternative methods which can freeze molecular movements in solutions (or in gases) and immobilize them according to design. The applications of this method are far more versatile than was previously envisaged, as described in Chapters 7 and 8. The use of biomolecules as templates is also promising in molecular biology and pharmocology. We see many scientific papers in which the concept of molecular imprinting is efficiently utilized although it is not clearly described as such. No one doubts that this method is one of the keys in the 21st century.

References

1 G. Wulff, R. Grobe-Einsler, A. Sarhan, *Makromol. Chem.*, **1977**, *178*, 2817.
2 K. J. Shea, T. K. Doughertly, *J. Am. Chem. Soc.*, **1986**, *108*, 1091.
3 R. Arshady, K. Mosbach, *Macromol. Chem.*, **1981**, *182*, 687.
4 G. Vlatakis, L. I. Andersson, R. Muller, K. Mosbach, *Nature*, **1993**, *361*, 645.
5 M. J. Whitcombe, M. E. Rodriguez, P. Villar, E. N. Vulfson, *J. Am. Chem. Soc.*, **1995**, *117*, 7105.

Chapter 3
Experimental Methods (1) – Procedures of Molecular Imprinting

3.1
Introduction

In Chapter 2, basic concepts and principles of molecular imprinting have been presented. The superiority and high potential of this method should now be clear. In the following two chapters, we will discuss further details of this method. The focus of this chapter is »experimental procedures of molecular imprinting«. Main concerns are: (i) What kinds of reagents are necessary? (ii) What reaction conditions are appropriate? and (iii) What factors lead to high imprinting efficiency? Furthermore, some experimental examples are presented so that the reader can use the molecular imprinting method for his or her own experiments without difficulty. Another important aspect of this method, »precise evaluation of imprinting efficiency«, will be described in Chapter 4.

3.2
Reagents and Experimental Procedures

All the chemicals we need are (1) functional monomers carrying templates (either covalently or non-covalently), (2) crosslinking agents, (3) solvents for the polymerization, and (4) solvents (or bond-cleaving

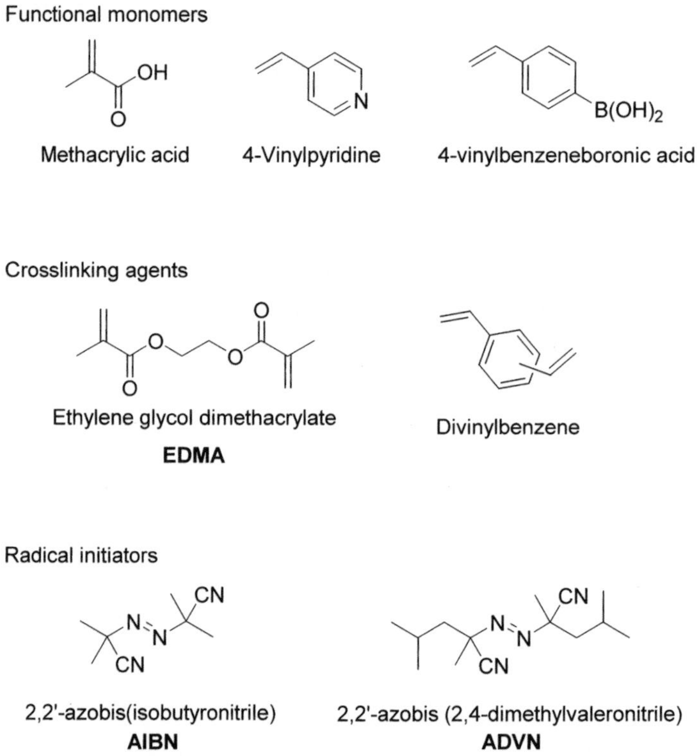
Functional monomers
OH
O
N
B(OH)2
Methacrylic acid
4-Vinylpyridine
4-vinylbenzeneboronic acid
Crosslinking agents
O
O
O
O
Ethylene glycol dimethacrylate
EDMA
Divinylbenzene
Radical initiators
CN
N
N
CN
CN
N
N
CN
2,2'-azobis(isobutyronitrile)
AIBN
2,2'-azobis (2,4-dimethylvaleronitrile)
ADVN

Fig. 3-1 Typical reagents

agents) to remove the templates from the polymers. Chemical structures of typical reagents used are presented in Fig. 3.1.

3.2.1
Functional Monomers

All kinds of polymerization (radical, anion, cation, and condensation) can be employed for molecular imprinting. The only requisite is that the polymerization can satisfactorily occur under the conditions where all the components (the templates, the crosslinking agents, non-covalent adducts between the monomer and the template in non-covalent imprinting, and others) remain intact. However, radical polymerization is most commonly used, mainly because of its versatile applicability and experimental easiness (note that radical polymerization is also most widely used in industry, mainly for economic reasons). In covalent imprinting, templates are bound to vinyl moieties by covalent linkages. Esters and amides of acrylic acid or methacrylic acid are most often used. The synthesis is easy in most cases. For non-covalent imprinting, vinyl monomers bearing appropriate functional groups are designed and synthesized. Furthermore, many functional monomers are also commercially available. However, it should be noted that these commercial monomers usually contain inhibitors or stabilizers (e.g., hydroquinones and phenols) to avoid undesired polymerization during their storage. Thus, commercial monomers must be distilled before use in your molecular imprinting experiments.

3.2.2 Crosslinking Agents

For molecular imprinting in organic solvents, ethylene glycol dimethacrylate (EDMA) and divinylbenzene are often used. A typical water-soluble crosslinking agent is *N,N'*-methylenebisacrylamide. The fundamental role of these reagents is to fix the guest-binding sites firmly in the desired structure. They also make the imprinted polymers insoluble in solvents and facilitate their practical applications. By using different kind of crosslinking agent, we can control both the structure of the guest-binding sites and the chemical environments around them.

For efficient imprinting, the reactivity of the crosslinking agent should be similar to that of the functional monomer (otherwise, either the functional monomer or the crosslinking agent polymerizes predominantly, and copolymerization cannot take place sufficiently). By choosing an appropriate crosslinking agent, random copolymerization occurs successfully, and the functional residues (derived from the functional monomers) are uniformly distributed in the polymer network.

The mole ratios of crosslinking agent to functional monomer are also important. If the ratios are too small, the guest-binding sites are located so closely to each other that they cannot work independently. In extreme cases, the guest binding by one site completely inhibits the guest binding by the neighboring sites. At extremely large mole ratios, however, the imprinting efficiency is damaged, especially when the crosslinking agents show non-covalent interactions with functional monomers and/or templates.

3.2.3 Solvents

The trivial role of solvents is to dissolve the agents for polymerization. However, they have more crucial roles. One of them is to provide porous structures to imprinted polymers, and promote their rates of guest binding. Release of the bound guest from the polymer is also facilitated by the porosity. In the polymerization, solvent molecules are incorporated inside the polymers and are removed in the post-treatment. During these processes, the space originally occupied by the solvent molecules is left as pores in the polymers. Polymers prepared in the absence of solvents are consistently too firm and dense, and hardly bind guests. Another role of solvents is to disperse the heat of reaction generated on polymerization. Otherwise, the temperature of reaction mixture is locally elevated, and undesired side-reactions occur there. Furthermore, the formation of monomer-template adduct, which is required for efficient non-covalent imprinting, is suppressed.

Choice of solvents is dependent on the kind of imprinting. In covalent imprinting, many kinds of solvents are employable as long as they satisfactorily dissolve all the components. In non-covalent imprinting, the choice of solvent is more critical to the promotion of the formation of non-covalent adducts between the functional monomer and the template and thus to enhancement of the imprinting efficiency. Chloroform is one of the most widely used solvents, since it satisfactorily dissolves many monomers and templates and hardly suppresses hydrogen bonding. However, commercially available chloroform is usually stabilized by ethanol to avoid the formation of poisonous phosgene during the storage. This ethanol is inappropriate for most molecular imprinting (especially for non-covalent imprinting), since it inhibits hydrogen bonding between monomer and template. In order to obtain good results, commercial chloroform must be distilled before use to remove ethanol. Carbon tetrachloride is not appropriate for molecular imprinting (with a few exceptions). In radical polymerization, it is a chain-transfer agent and decreases the molecular weight of polymers.

3.2.4 Polymerization Procedures

Radical polymerization can be initiated by using thermal decomposition of radical initiators. Typically, 2,2′-azobis(isobutyronitrile) (AIBN) and 2,2'-azobis(2,4-dimethylvaleronitrile) (ADVN) are used. The initiation radicals formed by the decomposition attack the monomer, producing the propagating radicals. The reactions are very simple and economical. However, it is important to remove molecular oxygen from polymerization mixtures, since it traps the radical and retards (or even stops) the polymerization. In order to remove oxygen, degassing with nitrogen or argon, as well as freeze-and-thaw cycles under reduced pressure, is effective.

In some cases, non-covalent adducts between functional monomer and template are too unstable to be used at higher temperatures, and the polymerization must be carried out at lower temperatures. Under these conditions, the thermal decomposition of initiator cannot be used to initiate the polymerization, and the initiators are decomposed with UV-light irradiation (photo-initiation never requires high temperatures). If the monomers themselves absorb UV light sufficiently, the polymerization is initiated even in the absence of any radical initiators.

3.3 Covalent Imprinting

One of the keys for successful covalent imprinting is the choice of the covalent linkage which connects a functional monomer with a template. These linkages must have both »stable« and »reversible« characters, which in a sense contradict each other. Thus, they must be sufficiently stable and be kept intact during the polymerization, but must be easily cleaved later (in step 3 in Fig. 2.1) under mild conditions without damaging the imprinting effects. In order to bind the target guest (and release it) promptly, both the formation and dissociation of the covalent linkage must be fast. However, the number of covalent bonds which fulfill both of these thermodynamic and dynamic requirements is small. The linkages available at present are boronic acid esters, acetals, ketals, Schiff bases, disulfide bonds, coordination bonds, and some others. The experimental procedures for the imprinting are essentially the same for all cases.

3.3.1 Imprinting with Boronic Acid Esters

Boronic acid esters are synthesized from boronic acid and *cis*-1,2- or *cis*-1,3-diol compounds (vinyl groups for the polymerization are tethered to the organic portions, as described in Example 3.1 below). Its formation and dissociation are fast and easy. They have five-membered cyclic structures, which are rigid enough to fix the covalent conjugates in a desired conformation. Thus, effective molecular imprinting is accomplished. After the polymerization, these linkages are cleaved by hydrolysis, and the boronic groups in the conjugates are arranged suitably for guest binding. These conjugates are especially useful for molecular imprinting toward carbohydrates and their derivatives which have *cis*-diol moieties. As expected, the imprinting is still more effective when carbohydrate templates have two or more *cis*-diol structures.

Example 3.1: Receptors toward saccharides (see Fig. 2.2)
4-Nitrophenyl α-D-mannopyranoside-2,3;4,6-di-*O*-(4-vinylphenylboronate) was prepared by distilling azeotropically the benzene solution of 4-nitrophenyl mannopyranoside and tris(4-vinylphenyl)boroxine. The mixture of this boronic acid ester (3.0 g, 5.7 mmol), methyl methacrylate (4.5 g, 45 mmol), ethylene glycol dimethacrylate (7.5 g, 38 mmol), and AIBN (0.1 g, 0.6 mmol) in acetonitrile (15 mL) was degassed three times, sealed under argon, and polymerized at 80 °C for 2 days. The obtained polymer was milled, extracted with dry acetonitrile, and dried at 40 °C *in vacuo*. The boronic acid ester linkages were cleaved by treating the polymer with 1:1 ethanol/water mixture for three days. If particles of appropriate size are necessary, the polymer should be sieved accordingly [1].

Example 3.2: Receptors toward sialic acid (see Fig. 3.2)
By removing water under a reduced pressure at 40 °C, the ester of 4-vinylphenylboronic acid (0.30 g, 2.0 mmol) and sialic acid (0.62 g, 2.0 mmol) were prepared in dry pyridine (200 mL). The product was directly used for the polymerization (the covalent conjugate did not have to be isolated, since its formation was almost quantitative). This ester, ethylene glycol dimethacrylate (5.6 g, 28.2 mmol) and ADVN were dissolved in DMF (3.2 mL), and the polymerization was initiated by irradiating with UV light at 4 °C. After 12 h, the polymer was ground, and treated with 1:1 mixture of 0.01 M hydrochloric acid and methanol. By this post-treatment, the boronic acid esters were hydrolyzed and the sialic acids were removed from the polymer [2].

3.3.2
Imprinting with Carbonate Esters

Carbonate esters are less stable than carboxylic acid esters, and can be completely hydrolyzed by sodium hydroxide solution (e.g., 1 M) in methanol. In the molecularly imprinted polymers, OH residues (of the phenol in Example 3.3) are placed at the appropriate position in the binding sites.

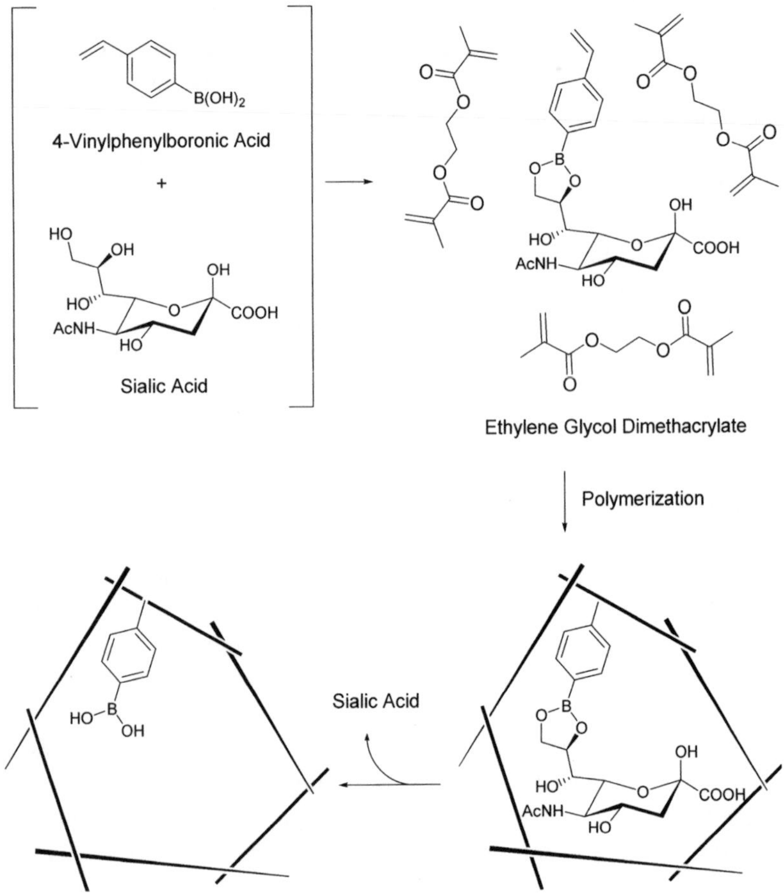

Fig. 3-2 Molecular imprinting of sialic acid using 4-vinylphenylboronic acid as a functional monomer

In a classical sense, the guest binding by these imprinted polymers should occur through formation of the original carbonate ester. However, this guest binding mode is not usually employed, since it is too slow, and, still more critically, it requires an activated carbonate group (e.g., Cl-C(O)-O-) in the guest compound used. This imposes a strict restriction in the application. Accordingly, it is more common for the guest binding to use hydrogen bonding between the guest (cholesterol in Example 3.3) and the OH of the polymer which is formed by hydrolyzing the covalent conjugate in the post-treatment. This OH (polymer-OH) takes an appropriate position for the hydrogen bonding, since the space occupied by the carbonyl group in the covalent conjugate (monomer-O-**C(O)**-O-template) is now allotted to the **H** atom for the hydrogen bonding (polymer-O-**H**· · · O(H)-guest). The guest binding is sufficiently fast, as is the case in non-covalent imprinting (see Chapter 2).

Example 3.3: Receptors toward cholesterol (see Fig. 2.4)
Cholesteryl 4-vinylphenyl carbonate was synthesized by reacting 4-vinylphenol (obtained by hydrolyzing 4-acetoxystyrene) and cholesteryl chloroformate (commercially available) in dry THF. To the mixture of this carbonate ester (5 mol%) and ethylene glycol dimethacrylate (95 mol%) in hexane (2 mL/g, solvent/monomers), was added AIBN (1 mol% with respect to the C=C bonds). The solution was degassed and sealed at a reduced pressure. The polymerization was carried out at 65 °C for 24 h. The resulting polymer was washed with methanol, ground (average size: 30 μm), extracted with methanol in a Soxhlet apparatus for 12–18 h, and dried *in vacuo*. In order to hydrolyze the carbonate esters, the polymer particles were suspended in 1 M NaOH/methanol. After the suspension was refluxed for 6 h, it was added to dilute hydrochloric acid and filtered. The polymer was further washed with water, methanol, and ether, extracted with methanol and then with hexane, and finally dried *in vacuo* [3].

3.3.3 Imprinting with Acetals and Ketals

Ketone and aldehyde compounds are reacted with 1,3-diol compounds, and the resultant ketals and acetals products are used as functional monomers. At the guest-binding sites in the imprinted polymers, these 1,3-diol groups are placed complementarily to the guest.

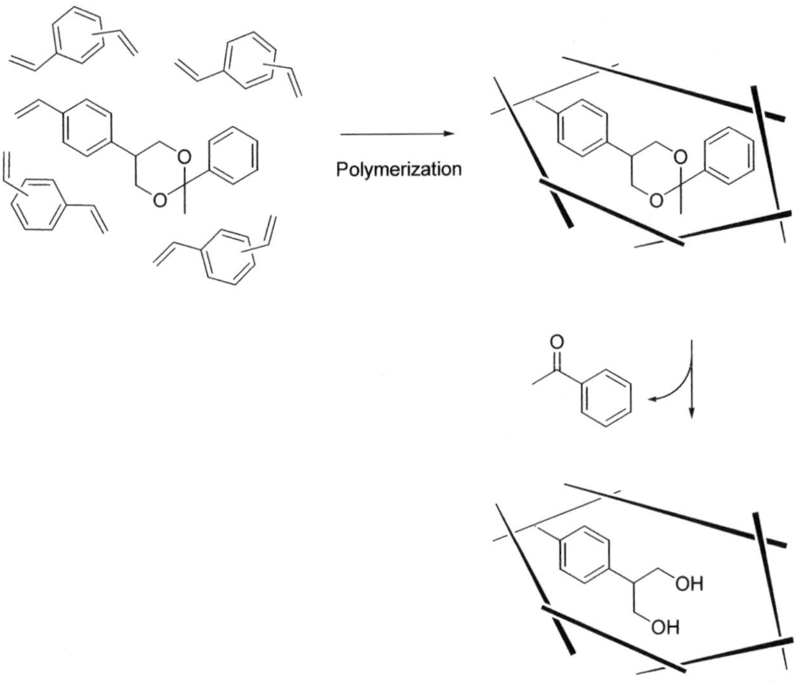

Polymerization

Fig. 3-3 Molecular imprinting using ketal as a functional monomer

Example 3.4: Receptors toward ketones (see Fig. 3.3)
From 2-(*p*-vinylphenyl)-1,3-propanediol and acetophenone (the template), the corresponding ketal was prepared. This ketal conjugate (1–2 mol%) was copolymerized with divinylbenzene with AIBN as the initiator (acetonitrile was used as diluent). The template was removed by hydrolyzing the ketal with methanol/H_2O/H_2SO_4.

3.3.4
Imprinting with Schiff Bases

Aldehyde compounds are reacted with amino compounds to yield Schiff base compounds. These reactions are easy to achieve and proceed almost quantitatively. Aldehyde groups can be placed in the guest-binding sites to bind amino compounds. Alternatively, amino groups in the binding sites can be used to bind aldehyde compounds.

Example 3.5: Receptors toward amines (see Fig. 3.4)

The mixture of *N*-(5-vinylsalicylidene)-L-phenylalanine methyl ester (3.2 mol%) and divinylbenzene (96.8 mol%; 70 wt% of *m*- and *p*-isomeric mixture) was dissolved in acetonitrile/benzene (1:1, v/v; 1 mL per g of monomer mixture). The solution was polymerized with AIBN (0.5 mol%) for 20 h at 75 °C. The resulting polymer was ground and the polymer particles obtained were treated with CH_3OH/1 M HCl (4:1,

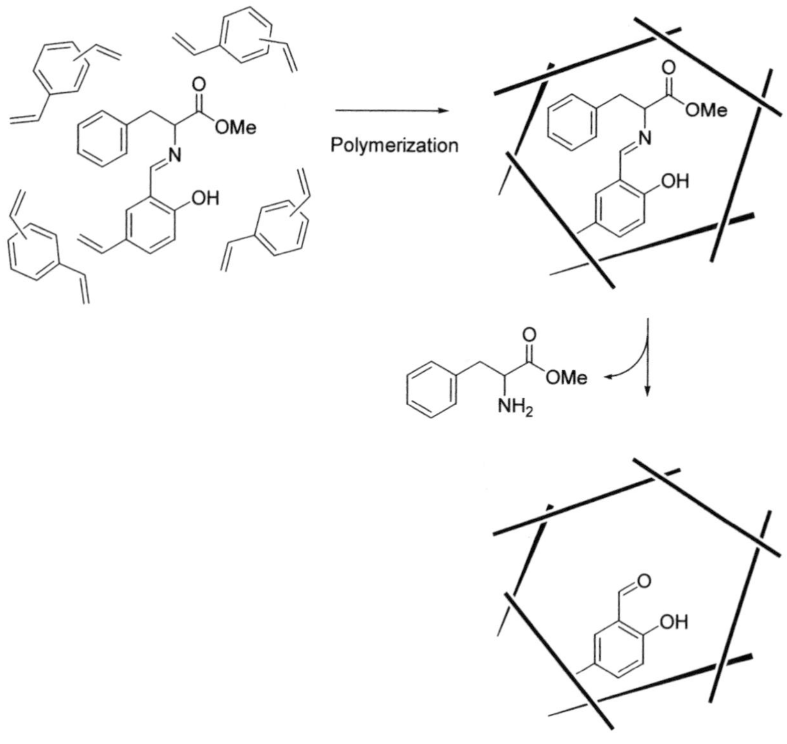

Fig. 3-4 Molecular imprinting using Schiff base as a functional monomer

v/v; 200 mg polymer /10 mL) for 15 h at room temperature to cleave the azomethine bond [4].

3.3.5 Imprinting with S–S Bonds

These bonds are sufficiently stable under polymerization conditions, but are easily cleaved by reduction. The –SH groups in the imprinted polymers bind the guest. One of the interesting characteristics is a strong discrimination between the guests bearing –OH groups and the ones bearing –SH groups. The hydrogen bonding between –SH and HO– is far stronger than that between –SH and HS–. Thus, the imprinted polymers present overwhelmingly prefer the OH-bearing guests to the corresponding SH-bearing ones (e.g., the polymer in Example 3.6 strongly binds phenol but hardly binds thiophenol). Difference in atomic size between O and S is also responsible for this notable recognition. The guest-binding site is sterically restrained so that the difference in atomic size is critical.

Example 3.6: Receptors toward phenol (see Fig. 3.5)
A mixture of allyl phenyl disulfide (0.94 g, 5.2 mmol), divinylbenzene (mixture of *m*- and *p*-isomers; 12.4 g, 95 mmol), and AIBN (0.36 g, 2.2 mmol) in chloroform (10 mL) was degassed with nitrogen for 5 min, and was polymerized by UV light irradiation for 24 h at 5 °C and for a further 3 h at 80 °C. The polymer was ground, and the particles were suspended in methanol. The S–S linkages were cleaved by the reduction with $NaBH_4$ for 12 h. This cleavage procedure was repeated twice to complete the reaction [5].

3.3.6 Imprinting with Coordination Bonds

Some coordination bonds between metal ions and ligands are sufficiently stable for covalent imprinting. Vinyl groups are attached to metal complexes, and these »polymerizable« metal complexes are used as

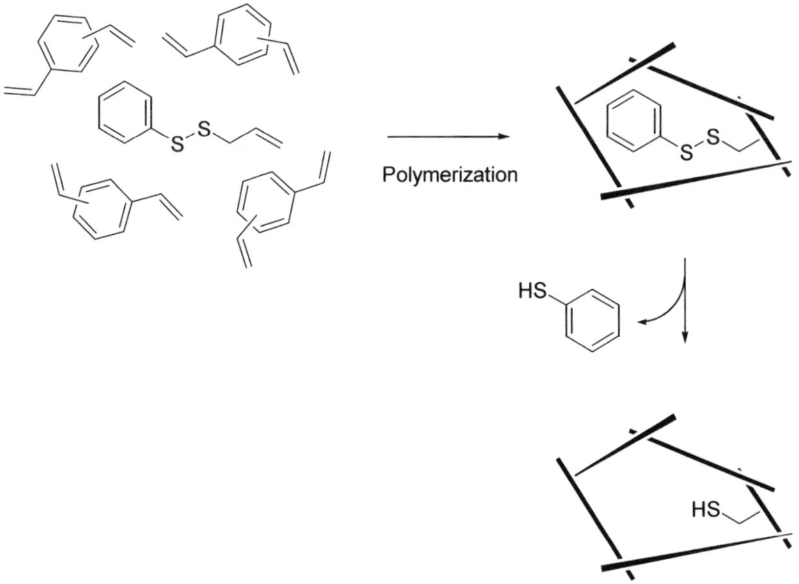

Polymerization

Fig. 3-5 Molecular imprinting using allyl phenyl disulfide as a functional monomer

functional monomers. These are polymerized in the presence of an appropriate ligand (template), and the whole structures of metal-ligand complexes are frozen in the polymer. After the ligand is removed, the guest binding occurs via formation of the same coordination bond. In Example 3.7, cinchonidine (the template) is coordinated to zinc(II)-porphyrin in the imprinted polymer. Interestingly, the intensity of fluorescence from the Zn(II) complex diminishes when the guest is bound to the Zn(II) ion, implying a potential application to a sensor.

Example 3.7: Receptors toward cinchonidine (see Fig. 3.6)
To a solution of cinchonidine (0.22 mmol) in chloroform (7.8 mL) were added 5,10,15-tris(4-isopropylphenyl)-20-(4-methacryloyloxyphenyl)porphyrin zinc(II) complex (0.22 mmol), methacrylic acid (0.45 mmol), ethylene glycol dimethacrylate (15 mmol), and ADVN (16 mg). After

degassing with nitrogen, the mixture was polymerized at 45 °C for 12 h in the dark. The polymer was ground and washed with methanol/acetic acid (7/3, v/v) to remove the cinchonidine [6].

Furthermore, this methodology can provide highly sophisticated polymeric materials, in which unique functions of metal complexes (catalysis, electron-transfer, and others) are combined with the molecule-recognizing activity of the polymer. For example, aldolase-mimic

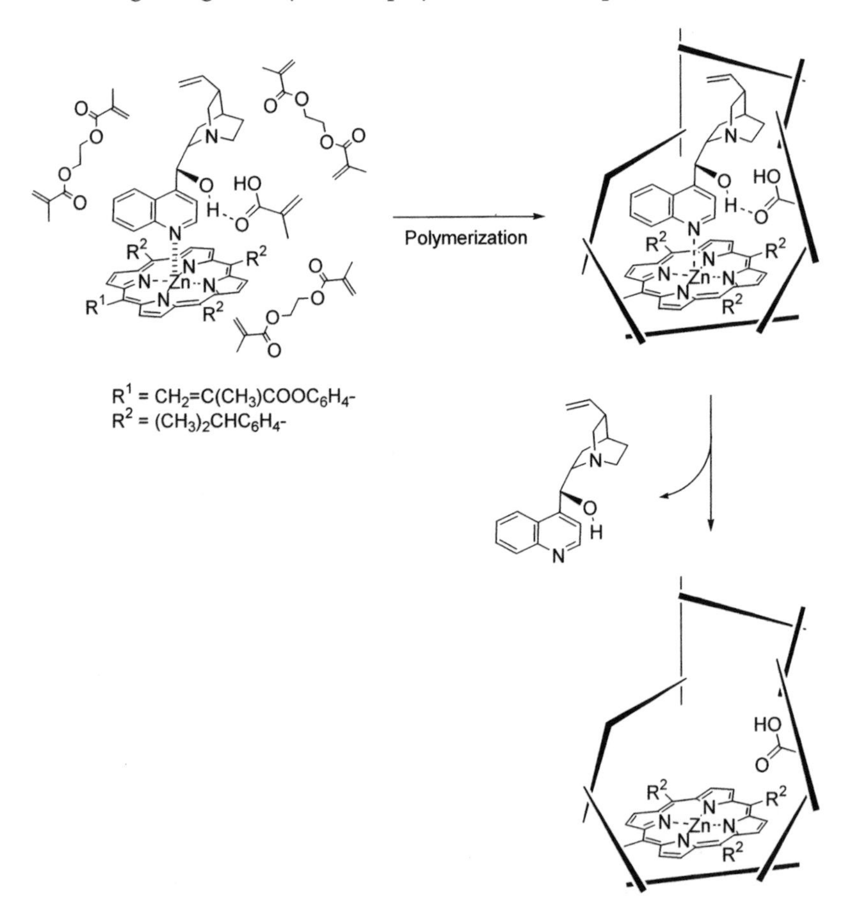

Fig. 3-6 Molecular imprinting of cinchonidine using zinc(II)-porphyrin complex and methacrylic acid as functional monomers

was prepared by imprinting 4-vinylpyridine toward the Co(II) complex of dibenzoylmethane (Example 3.8). This template resembles the transition-state for the Co(II)-catalyzed cross-aldol condensation between acetophenone and benzaldehyde. Thus, this imprinted polymer stabilized the transition state to a greater extent than it stabilized the initial state. As a result, the difference in free energy between these two states is decreased, and efficient catalysis of the condensation reaction is achieved. An artificial »catalytic antibody« has been obtained (a catalytic antibody is an antibody obtained toward the transition state analog of a reaction, and catalyzes the corresponding reaction).

Example 3.8: Preparation of artificial catalytic antibody (see Fig. 3.7)
Vinylpyridine (420 mg), styrene (4.20 g), divinylbenzene (5.20 g), dibenzoylmethane (448 mg; a transition-state analog), cobalt(II) acetate (498 mg), and ADVN (100 mg) were dissolved in anhydrous methanol (2.5 mL) and chloroform (6.7 mL). The resulting mixture was sonicated under vacuum and purged with N_2 for 5 min at 0 °C. The polymerization was carried out at 45 °C for 24 h. The resulting polymer was ground and sieved. The polymer particles were packed into a stainless steel column and washed with methanol/acetic acid (7/3) and then with methanol to remove dibenzoylmethane and cobalt(II) acetate.

Then the polymer was incubated with cobalt(II) ion again. When acetophenone and benzaldehyde were added, they were placed in the cavity appropriate for the reaction due to the imprinting effect (note that the template is similar to the transition state of the reaction). The cobalt(II)-catalyzed aldol reaction occurred smoothly [7].

3.4 Non-covalent Imprinting

The reaction procedures for non-covalent imprinting are far simpler than those for covalent imprinting. Functional monomers are simply combined with template in the polymerization mixtures and copolymerized with crosslinking agent. The adducts between the functional

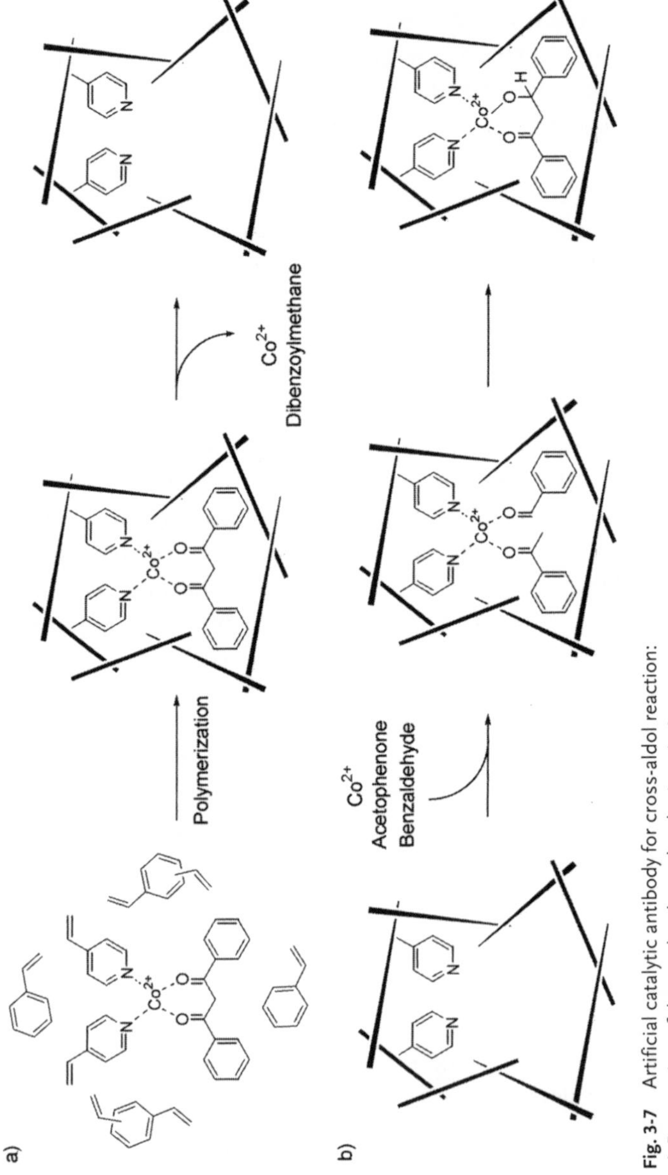

Fig. 3-7 Artificial catalytic antibody for cross-aldol reaction:
a) Preparation of the catalyst by molecular imprinting;
b) Cross-aldol reaction of acetophenone with benzaldehyde

monomer(s) and the template are formed *in situ* by non-covalent interactions, and are frozen in the polymers. There is no need for the synthesis of covalent conjugates prior to polymerization, as is the case in covalent imprinting. Furthermore, the template is easily removed after the polymerization, by simple extraction. Many of practically important molecules (e.g., pharmaceuticals, herbicides, biologically active substances, and environmental contaminants) possess polar groups such as hydroxyl, carboxyl, amino, and amide for the required non-covalent interactions. Because of its simplicity and versatility, non-covalent molecular imprinting has been widely attempted (the shortcomings of this type of imprinting are described in the previous chapter).

In principle, any kind of non-covalent interactions should be effective and employable for the imprinting. However, hydrogen bonding is most appropriate for precise molecular recognition, since this bonding is highly dependent on both distance and direction between monomers and templates. Accordingly, various monomers which bear the required functional groups (e.g., carboxyl, amino, pyridine, hydroxyl, and amide groups) complementarily to the template are chosen. It should be noted that the nature of non-covalent interactions (the contributions of hydrogen bonding and electrostatic interaction) is governed by the p*K* values of both components. When both the acid and the base are very strong, for example, a proton is completely transferred from the acid to the base, and thus their adducts are formed mostly through electrostatic interaction. This is unfavorable for precise molecular recognition, since electrostatic interaction is less dependent on distance and direction (its strength is determined only by the distance between the components). With the combination of acid and base of intermediate strength, however, the proton exists between the acid and the base, and thus the hydrogen-bonding nature is dominant. It is under these conditions that efficient imprinting should be achieved. If both the acid and the base are weak, the interaction itself is too weak and is inappropriate for the imprinting.

3.4.1
Carboxyl Groups as Hydrogen-bonding Site

In aprotic solvents, the carboxylic acid residues in methacrylic acid and acrylic acid form hydrogen bonds with various basic templates and form non-covalent adducts. These two monomers are being widely used in the current polymer industry, and are available on quite a large scale at low cost. Molecular imprinting with the use of hydrogen bonding between methacrylic acid and atrazine (a herbicide) is described in detail in Chapter 6.

3.4.2
Pyridines as Hydrogen-bonding Site

Pyridine is intrinsically a rather weak base (pK_a = 5.4), but forms sufficiently strong hydrogen-bonding adducts with various acidic compounds. As functional monomers for molecular imprinting, 2-vinylpyridine and 4-vinylpyridine are commercially available.

Example 3.9: Receptors toward phenoxyacetic acid (see Fig. 3.8)
A mixture of 2,4-dichlorophenoxyacetic acid (2,4-D: 1 mmol), 4-vinylpyridine (4 mmol), ethylene glycol dimethacrylate (20 mmol), and ADVN (0.31 mmol) in a methanol/water mixture (4 mL/1 mL) was sonicated and purged with nitrogen for 2 min. The solution was polymerized at 45 °C for 4 h, and then at 60 °C for 2 h. The polymer was ground and sieved in acetone. The particles were washed by methanol/acetic acid (7/3, v/v; twice), acetonitrile/acetic acid (9/1, v/v; twice), acetonitrile (once), and methanol (twice) for 2 h each time, followed by centrifugation. The particles were then suspended in acetone for 4 h. The particles in suspension were collected and the procedure was repeated four times. The solvent was removed by centrifugation and the particles were dried *in vacuo* [8].

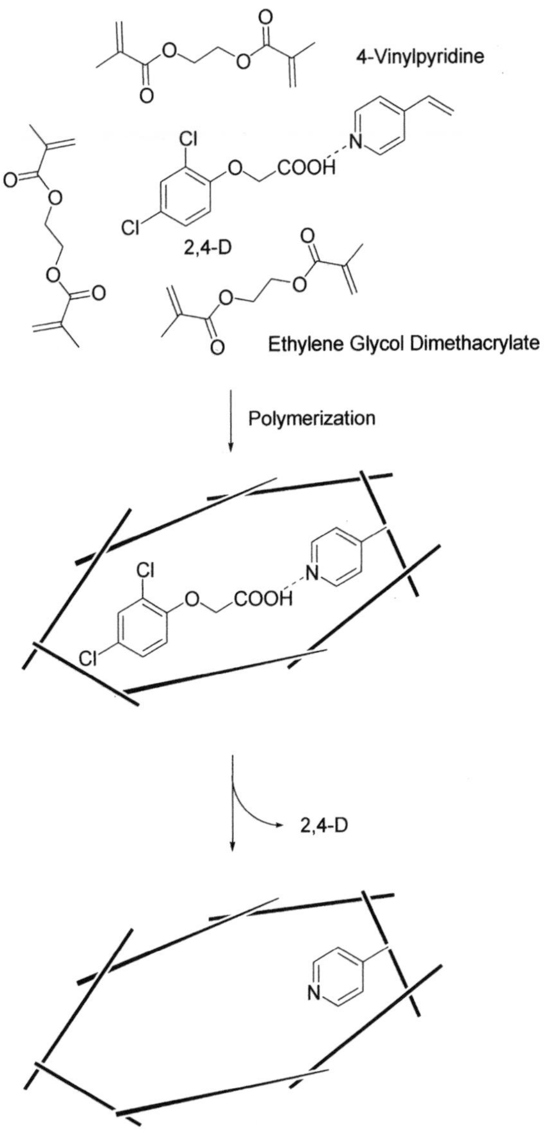

Fig. 3-8 Molecular imprinting of 2,4-dichlorophenoxyacetic acid (2,4-D) using 4-vinylpyridine as a functional monomer

3.4.3
Guest-binding Sites Having Multiple Hydrogen-bonding Sites

When two or more hydrogen bonds are simultaneously formed between functional monomer and template, the resultant adducts become extremely stable, and thus the molecular imprinting is still more efficient. 2,6-Bis(acrylamido)pyridine is one of the most useful monomers for this purpose, since it has two hydrogen-bonding donors (the amide groups) and one hydrogen-bonding acceptor (the nitrogen atom of pyridine) in one molecule. With barbiturates, for example, it forms a 2:1 complex by using six hydrogen bonds. As expected, the molecular imprinting is highly efficient.

Example 3.10: Receptors towards barbiturates (see Fig. 3.9)
A mixture of 2,6-bis(acrylamido)pyridine (1 mmol), cyclobarbital (2 mmol), ethylene glycol dimethacrylate (20 mmol), and ADVN in chloroform was purged with N_2 gas. The polymerization was performed at 45 °C for 24 h and then at 90 °C for 3 h. The resulting polymer was ground and sieved to yield polymer particles (26–63 μm). The particles were packed into a stainless steel column and washed with methanol and chloroform to remove cyclobarbital [9].

3.4.4
Guest-binding Sites Having Pyridinium Salt for Electrostatic Interaction

The free energy of electrostatic interaction is simply proportional to the reciprocal of distance between the charges, and is independent of direction. This spherically symmetric potential is unfavorable for precise molecular recognition (compare with the strictness of hydrogen bonding with respect to direction and distance). However, electrostatic interaction is appropriate for the imprinting in polar solvents such as alcohol and water, since it can work satisfactorily therein. The selectivity can be improved by combining it with other interactions (e.g., hydrogen bonding and charge-transfer interactions).

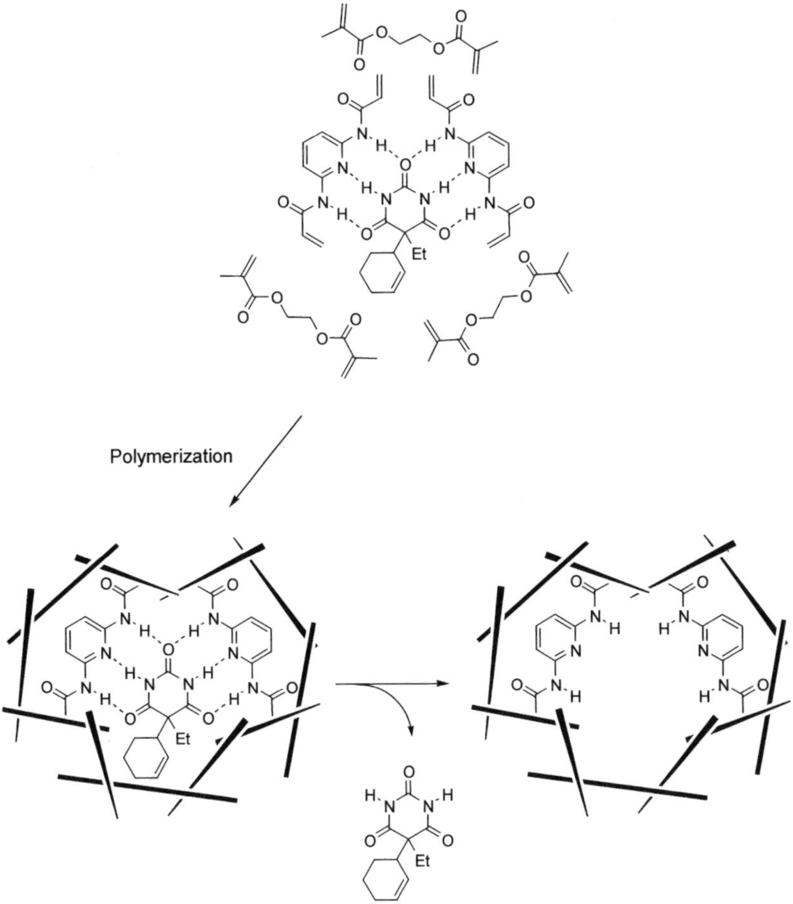

Fig. 3-9 Cyclobarbital-imprinting using 2,6-bis(acrylamido)pyridine

Example 3.11: Receptors toward cAMP (see Fig. 3.10)
To the solution of *trans*-[*p*-(*N*,*N*-dimethylamino)styryl]-*N*-vinylbenzylpyridinium chloride (31.5 mg, 0.084 mmol) and cAMP (41.7 mg, 0.12 mmol) in methanol (30 mL), 2-hydroxyethyl methacrylate (0.675 g, 5.2 mmol), trimethylolpropane trimethacrylate (29.0 g, 86 mmol), and AIBN (0.665 g, 4.1 mmol) were added. The resulting mixture was

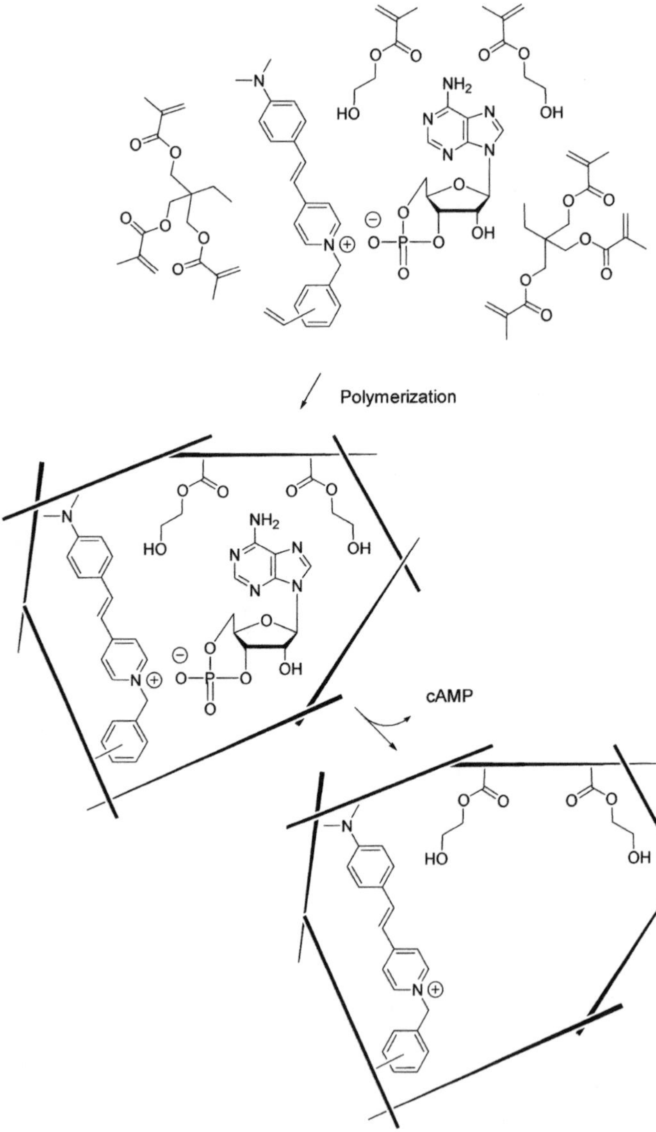
NH2
HO
OH
Polymerization
cAMP

Fig. 3-10 cAMP-imprinting by the use of the electrostatic interaction between pyridinium cation and phosphate anion

purged with nitrogen for 10 min, polymerized by irradiation (350 nm) at room temperature for 1.5 h, and then placed at 60 °C for 24 h. The resulting polymer was ground in a mortar and sieved (45–106 μm). The polymer particles were washed by 300 mL of water/methanol (7:3, v/v) and then 300 mL of methanol with a Soxhlet extractor [10].

3.4.5 Apolar Binding Sites

Many planar dye molecules are bound to double-stranded DNA by being accommodated between adjacent Watson-Crick base pairs. These non-covalent interactions (intercalation) are used for molecular imprinting [11]. Various hydrophobic interactions are also useful for the purpose [12].

3.5 »Dummy Molecular Imprinting«

This method is used to prepare artificial receptors toward certain bioactive compounds or environmental hormones (e.g., dioxin). When we wish to achieve molecular imprinting with these templates, it often happens that the template compound is not available in a sufficient amount. Alternatively, the template can be too toxic and dangerous to be used in laboratories. Under these conditions, »direct« molecular imprinting is hard to achieve. In these cases, an appropriate substitute compound which has a similar structure to the real template but is more easily available (or non-toxic) is used as the template. In the following example, an artificial receptor for atrazine (a herbicide) is prepared by using trialkylmelamine (non-toxic) as the template. These two chemicals resemble each other, so that the imprinted polymer obtained with the dummy binds atrazine selectively and effectively. This »dummy molecular imprinting« is an extension of non-covalent imprinting, and enormously widens the spectrum of target compounds.

Example 3.12: Imprinting with dummy template for recognition of atrazine (see Fig. 3.11)

The solution of triethylmelamine (0.35 g, 1.7 mmol), methacrylic acid (0.58 g, 6.7 mmol), ethylene glycol dimethacrylate (9.3 g, 47 mmol), and AIBN (0.12 mg, 0.73 mmol) in chloroform (25 mL) was sonicated and bubbled with nitrogen gas. The polymerization was carried out by UV light irradiation at 5 °C for 12 h. The obtained polymer was crushed and ground in a mortar, and was suspended in acetonitrile/chloroform (1/1,

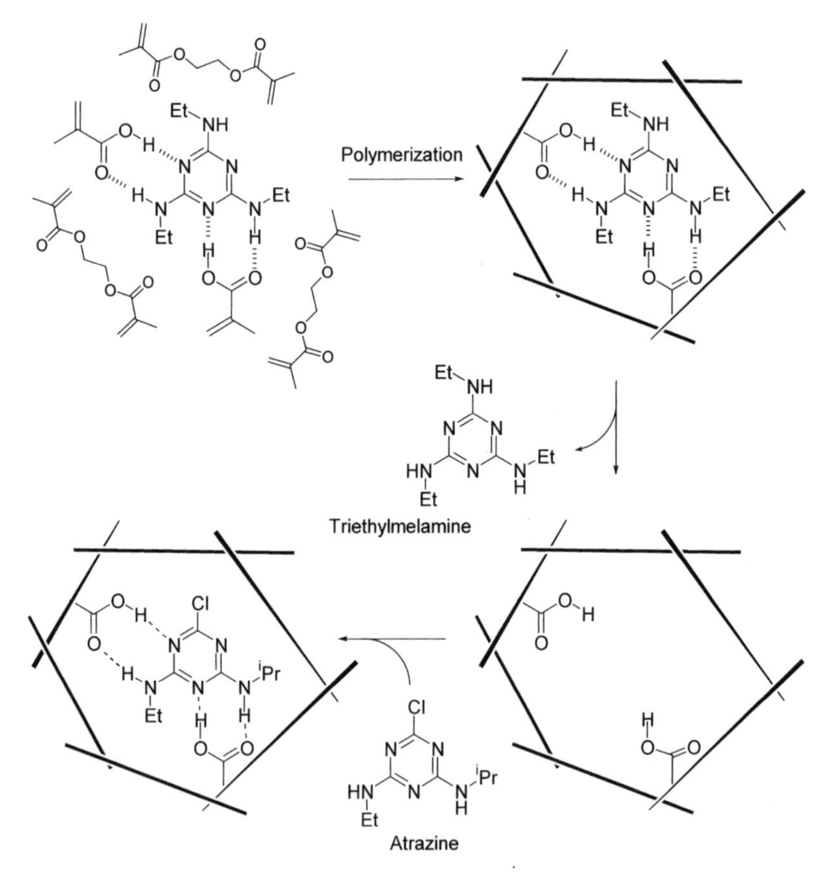

Fig. 3-11 Molecular imprinting by triethylmelamine as a dummy template for atrazine

v/v). The polymer particles were packed in a stainless steel column, washed with 300 mL of methanol/acetic acid (7/3, v/v), and used for column chromatography [13].

References

1 G. Wulff, W. Vesper, R. Grobe-Einsler, A. Sarhan, *Makromol. Chem.*, **178**, 2799 (1977).
2 A. Kugimiya, J. Matsui, T. Takeuchi, K.Yano, H. Muguruma, A. V. Elgersma, I. Karube, *Anal. Lett.*, **28**, *2317 (1995).*
3 *M. J. Whitcombe, M. E. Rodriguez, P. Villar, E. N. Vulfson, J. Am. Chem. Soc.*, **117**, 7105 (1995).
4 G. Wulff, J. Vietmeier, *Makromol. Chem.*, **190**, 1717 (1989).
5 T. Mukawa, T. Goto, H. Nariai, Y. Aoki, A. Imamura, T. Takeuchi, *J. Pharm. Biomed. Anal.*, in press.
6 T. Takeuchi, T. Mukawa, J. Matsui, M. Higashi, K. D. Shimizu, *Anal. Chem.*, **73**, 3869 (2001).
7 J. Matsui, I. A. Nicholls, I. Karube, K. Mosbach, *J. Org. Chem.*, **61**, 5414 (1996).
8 K. Haupt, A. Dzgoev, K. Mosbach, *Anal. Chem.*, **70**, 628 (1998).
9 K. Tanabe, T. Takeuchi, J. Matsui, K. Ikebukuro, K. Yano, I. Karube, *J. Chem. Soc., Chem. Commun.*, **1995**, 2303.
10 P. Turkewitsch, B. Wandelt, G. D. Darling, W. S. Powell, *Anal. Chem.*, **70**, 2025 (1998).
11 H. Bünemann, N. Dattagupta, H. J. Schuetz, W. Müller, *Biochemistry*, **20**, 2864–2874 (1981).
12 (a) G. Wulff, A. Sarhan, J. Gimpel, E. Lohmar, *Chem. Ber.*, **107**, 3364–3376 (1974). (b) G. Wulff and E. Lohmar, *Isr. J. Chem.*, **18**, 279 (1979).
13 J. Matsui, K. Fujiwara, T. Takeuchi, *Anal. Chem.*, **72**, 1810 (2000).

Chapter 4
Experimental Methods (2) – Evaluation of Imprinting Efficiency

4.1
Introduction

In the previous chapter, we learned how to achieve molecular imprinting reactions. Simply by polymerizing appropriate functional monomers in the presence of templates, polymeric receptors toward the target compound can be obtained. The next essential step is the evaluation of imprinting efficiency (whether or not our imprinted polymer adequately and accurately remembers the template). Other concerns are (i) how efficient is the guest binding and (ii) how strict is the discrimination between the target guest compound and the others. On the basis of the data on these points, we can improve the reaction conditions for molecular imprinting (choice of functional monomers, crosslinking agents, polymerization solvents, polymerization temperature, etc.) and hence prepare polymeric receptors possessing still better properties and functions.

Experimentally, the guest-binding activity of the imprinted polymer is measured by either chromatographic experiments or batchwise guest-binding experiments. By comparing the activity of our imprinted polymer with the corresponding activity for a non-imprinted polymer (prepared in the absence of template), the magnitude of the molecular imprinting effect is evaluated.

4.2 Chromatographic Experiments

Imprinted polymers are packed into stainless steel column-tubes and used as the stationary phase for high-performance liquid chromatography (HPLC). As the name indicates, the binding strength and binding selectivity toward guest compounds are analyzed in terms of their chromatographic behavior, since successful imprinting should selectively promote the binding toward the target guest and increase its retention time. The experimental procedures are simple and straightforward, and precise and comprehensive data are fairly easily obtainable (see Example 6.2 in Chapter 6 for details).

A typical HPLC chart is schematically illustrated in Fig. 4.1. The major of the two peaks corresponds to our target guest. Its retention time is t_g. The minor peak (the retention time = t_0) is that for the void marker (a standard), which is poorly bound by the polymer (e.g., acetic acid, acetone, or acetonitrile). An index of binding activity of the imprinted polymer toward our target guest, the capacity factor k, is defined by Eq. (1).

$$k = (t_g - t_0)/t_0 \tag{1}$$

In place of t_g and t_0, retention-volumes of guest and non-bound compound [$V(g)$ and $V(0)$] may be also adopted.

If our molecular imprinting is very successful, the k value of the target guest compound alone should be notably increased and those for all other compounds remain unchanged. This polymer clearly remembers

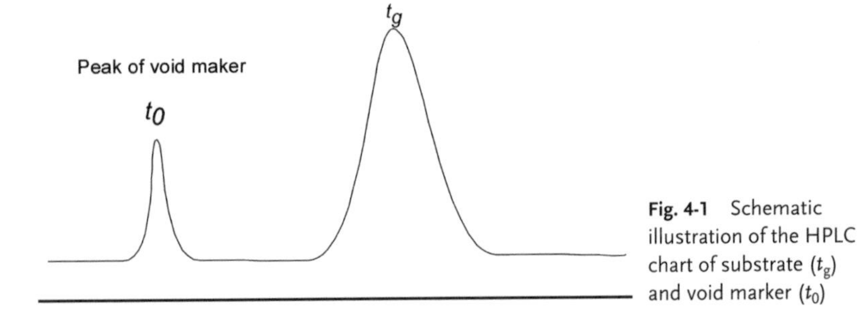

Fig. 4-1 Schematic illustration of the HPLC chart of substrate (t_g) and void marker (t_0)

Table 4-1 Capacity factors (*k*) of nucleotide analogs on HPLC columns packed with 9-ethyladenine-imprinted polymer of methacrylic acid and benzoic acid-imprinted polymer [1][a]

	Capacity factor k	
Substrate	9-Ethyladenine-imprinted	Benzoic acid-imprinted
9-Ethyladenine	54.8	1.7
1-Cyclohexyluracil	3.9	0.3
1-Propylcytosine	0.3	2.4
1-Propylthymine	0.4	0.2

a Mobile phase: acetonitrile/acetic acid/H_2O = 92.5/5/2.5

the structure of the template, and thus binds the target guest which is identical with (or similar to) the template. In unsuccessful imprinting, however, the difference between the *k* value of the target and the other *k* values would be marginal. Table 4.1 is a typical example. In the presence of 9-ethyladenine as the template, methacrylic acid is copolymerized with ethyleneglycol dimethacrylate in chloroform (»non-covalent imprinting«). The molecularly imprinted polymer shows a very large capacity factor *k* only for 9-ethyladenine. On the other hand, the *k* values of all other DNA-base derivatives are far smaller. Apparently, this imprinted polymer strictly recognizes the positions of hydrogen-bonding sites and their natures (either H-donor or H-acceptor) in the four kinds of DNA-bases. In its guest-binding sites, several carboxylic acids (derived from methacrylic acids) are placed complementarily to the adenine and bind this moiety through the formation of several hydrogen bonds. Consistently, guest specificity is almost nil when benzoic acid is used as the template for the imprinting. In this control experiment, the *k* values are quite small for all the four kinds of guests, since the template does not have hydrogen-bonding sites and accordingly the carboxylic acids, derived from methacrylic acid, are placed almost randomly in the polymer. Under these conditions, any satisfactory imprinting cannot be expected [1].

4.3 Batchwise Guest-Binding Experiments

In this method, the guest-binding activity of the imprinted polymer is directly determined in terms of the amount of guest bound by this polymer. First, a predetermined amount of polymer is added to guest solutions of varied concentrations. The polymer is usually insoluble in the solvent used. These mixtures are incubated for a sufficiently long period of time until the guest binding reaches the equilibrium. Then, the polymer is removed by centrifugation or filtration, and the concentration of guest in the liquid phase (C) is determined by HPLC, UV, or other analytical means. The guest selectivity is analyzed by comparing this value with the amount of guest (B_{bound}) bound by the polymer (per unit-weight) with those of other compounds.

4.4 Determination of Guest-Binding Constants

The dissociation constant K_D of the complex between guest and guest-binding site in the polymer is defined by Eq. (2).

$$\text{Guest} + \text{Binding Site} \overset{K_D}{\leftrightarrow} \text{Guest/Binding Site} \quad (2)$$

Here 1:1 stoichiometry of the complex is assumed. Under these conditions, the K_D is given by Eq. (3).

$$\begin{aligned} K_D &= (B_{unbound} \times C) \,/\, B_{bound} \\ &= (B_{max} - B_{bound}) \times C/B_{bound} \end{aligned} \quad (3),$$

where B_{max} is the maximal amount of guest bound by unit-weight of polymer (note that $B_{max} = B_{bound} + B_{unbound}$). The K_D has the dimension of concentration, and the guest binding is stronger when it is smaller. Accordingly, the plot of B_{bound} against C shows a gradual saturation (see Fig. 4.2a).

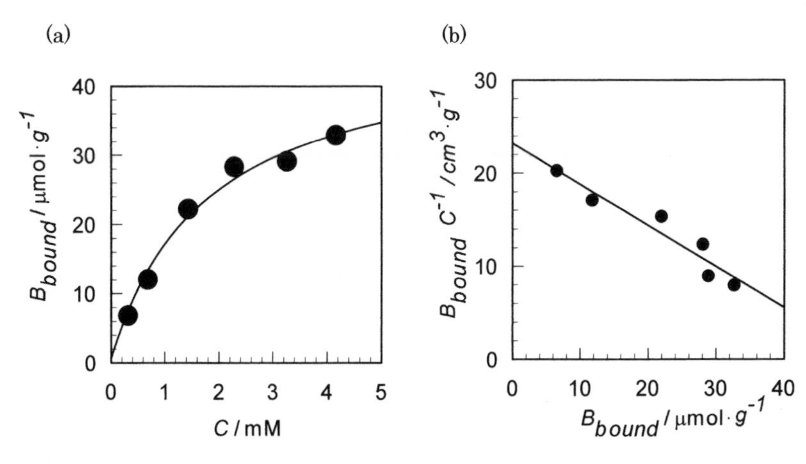

Fig. 4-2 Binding isotherm (a) and the corresponding Scatchard plot (b) for the binding of vancomycin (an antibiotic) by the vancomycin-imprinted cyclodextrin polymer [2]

Usually Eq. (3) is converted into Eq. (4) (Scatchard equation), and B_{bound}/C is plotted against B_{bound}.

$$\begin{aligned} B_{bound}/C &= (B_{max} - B_{bound})/K_D \\ &= -(1/K_D) \times B_{bound} + B_{max}/K_D \end{aligned} \tag{4}$$

From the slope and the intercept of the straight line obtained, the values of K_D and B_{max} are determined (see Fig. 4.2b). In ideal imprinted polymers, the K_D value should be small only for the target guest (strong and selective guest binding). The choice of C values employed is important in these experiments. If only too small C values are used for the plots, no saturation-phenomenon occurs. When all the C values are much larger than K_D, however, the guest-binding sites are almost completely occupied by the guest, irrespective of the C values. Both situations are unfavorable for the determination of K_D and B_{max}. In order to evaluate these values precisely, some of the C values should be below K_D and the others should be above it.

Example 4.1 Scatchard plots for the binding of vancomycin by the imprinted cyclodextrin (CyD) polymer in water.

Acryloyl-CyD (0.30 mmol), prepared by reacting CyD with *m*-nitrophenyl acrylate, and vancomycin (an antibiotic: 0.15 mmol) as the template are dissolved in 15 mL of Tris buffer solution ([Tris] = 5 mM, pH 8.0). After stirring for a few minutes, the polymerization is started by adding methylenebisacrylamide (3.0 mmol) and potassium persulfate (35 mg: radical initiator) under nitrogen at 50 °C. The system becomes opaque as the polymerization proceeds. After stirring for 2 h, the white precipitate is collected and washed with a large amount of hot water and acetone. Then these CyD-polymers are dried, ground with mortar and pestle, and are subjected to the following binding experiments.

The polymer containing 27 μmol of CyD residue (75 mg) is incubated in 3 mL of an aqueous solution of vancomycin (pH 8.0; 5 mM Tris buffer). After keeping the mixture at 5 °C for 24 h to complete the complex formation, the mixture is centrifuged and the concentration of the guest in the supernatant is analyzed by reversed-phase HPLC (Merck LiChrospher RP-18(e) ODS column). Figure 4.2a shows the isotherm (plots of B_{bound} as a function of C at a predetermined temperature). The isotherm gives an ideal saturation curve. The Scatchard plot from this isotherm is linear, as shown in Fig. 4.2b, giving the values of K_D and B_{max} as 1.6 mM and 44 μmol g^{-1}, respectively [2].

The data-analysis hitherto described is based on the assumption that all the guest-binding sites in an imprinted polymer are identical with each other. However, this is not necessarily the case. Rather, it often happens that two or more kinds of guest-binding sites are responsible for guest binding. There, Scatchard plots are not linear. A typical example of these non-linear plots, as well as the method for its treatment, is presented in Example 6.3 in Chapter 6.

References

1 D. Spivak et al., *J. Am. Chem. Soc.* **1997**, *119*, 4388–4393.

2 H. Asanuma et al., *Anal. Chim. Acta.*, **2001**, *435*, 25–33.

Chapter 5
Spectroscopic Anatomy of Molecular Imprinting Reactions

5.1
Introduction

The molecular imprinting method provides important polymeric receptors easily and cheaply. Unfortunately, however, mechanistic details of imprinting have not yet been satisfactorily clarified. One of the reasons is that this method has been developed primarily for practical applications, and thus improvements in binding activity and guest selectivity rather than an understanding of the mechanistic details of imprinting reactions have been the main concerns of many people. Furthermore, most imprinted polymers are insoluble in solvents, and the guest-binding sites are never identical with each other. These features have prevented detailed spectroscopic analysis. Yet the mechanistic information is crucially important if we wish to design efficient molecular imprinting systems. Accordingly, several attempts have been made to date, and fruitful and significant results have been obtained. This chapter deals with these spectroscopic results on pre-organized monomer-template adducts in the polymerization mixtures and guest-binding sites in the imprinted polymers.

5.2
Structures of Adducts in the Pre-organization Step

The first step in molecular imprinting is »pre-organization« of functional monomers with templates. In covalent imprinting, the conjugate is a well-characterized covalent molecule, so that this step is really simple. In non-covalent imprinting, however, the adducts are labile and dynamic, and thus efficient pre-organization is essential for successful molecular imprinting. The key questions are: (i) What is the structure of the non-covalent adduct in reaction mixtures? and (ii) What portion of the functional monomer is complexing with template under polymerization conditions? These properties should directly reflect the nature and efficiency of imprinting.

5.2.1
Adduct Formation in Solutions

In the analysis of these homogeneous specimens, ^{1}H-NMR is one of the most powerful tools. When a hydrogen bond is formed, for example, the electron density at hydrogen-bonding protons decreases. Accordingly, the NMR signal of this proton shifts toward lower magnetic field. The chemical shifts are also altered by the change in the chemical circumstance, anisotropic shielding by unsaturated compounds (e.g., aromatic rings), and many other factors. Information on the rate of ligand exchange is also obtained from the line-width of signal. In fast exchange, the signals are sharp, but they are broadened in slow exchange.

Example 5.1: Adduct formation for the imprinting of methacrylic acid (MAA) with L-phenylalanine anilide (PheNHPh) in acetonitrile.

Upon increasing MAA concentration, the ^{1}H-NMR signals for the amino protons of PheNHPh and the carboxyl group of MAA monotonously shift toward lower magnetic field. When the shift is plotted against the MAA/PheNHPh ratio, the first inflection point appears at MAA/PheNHPh = 1.0, corresponding to the formation of 1:1 adduct. However, these lower-field shifts never saturate there, and increase still

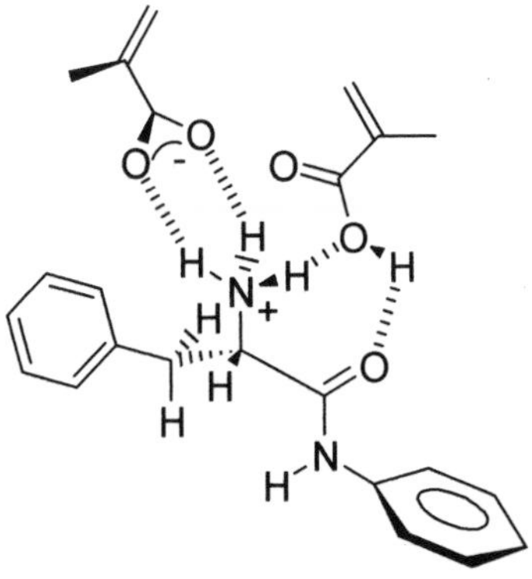

Figure 5-1 Pre-organized adduct between functional monomer (MAA) and the template (PheNHPh) [1]

more when $[MAA]_0$ is further increased. Apparently, two or more MAA molecules are interacting with one PheNHPh molecule. Consistently, the broadening of line-widths of these protons is maximized at around MAA/PheNHPh = 0.5, 3.5, and 6.0. One of the plausible structures of the 2:1 adduct is presented in Fig. 5.1: the first MAA molecule is interacting with the ammonium group of PheNHPh, and the second bridges this ammonium with the carbonyl group of the same PheNHPh molecule [1].

Adduct formation by other non-covalent interactions is also detectable by ^{1}H-NMR. When cholesterol (template) forms an inclusion complex with cyclodextrin (cyclic functional monomer composed of several glucose units), 18 methyl protons of the cholesterol show upfield shift due to the change in its chemical circumstance. The binding constant is 550 M^{-1}. This system provides ordered assemblies of two cyclodextrin molecules, which bind cholesterol cooperatively (see also Example 5.5 in this chapter) [2].

Infrared spectroscopy also reflects the structure of the pre-organized monomer-template adduct. When N–H, O–H, and C=O groups form hydrogen bonds with other hydrogen donors (or acceptors), these bonds are weakened, and accordingly their stretching vibration bands shift toward lower frequencies. In the infrared spectra, all of these bands are clearly distinguishable from other signals, and can be useful probes for the hydrogen bonding.

Example 5.2: Hydrogen bonding of 2,6-bis(acrylamido)pyridine.
This functional monomer, originally synthesized by Takeuchi et al. [3] involves both multiple hydrogen-bonding sites for the recognition of the guest and two vinyl groups for the crosslinking. Many more recognition sites can be introduced into the polymer, since this monomer does not require a conventional crosslinker.

In CH_2Cl_2, the N–H groups of this monomer are free from hydrogen bonding, and their stretching vibration bands are at around 3350 cm^{-1}. When a thymine derivative (the template) is added to this solution, these bands shift to 3210 cm^{-1}. Here, the amide groups form hydrogen bonds with the thymine (see Fig. 5.2), and thus their N–H bonds are weakened [4].

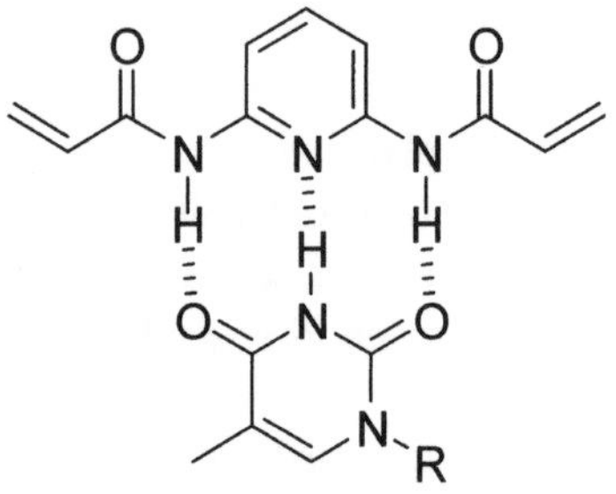

Figure 5-2 Multiple hydrogen bond formation between 2,6-bis(acrylamido)pyridine and thymine derivative

5.2.2
Adduct Formation on a Solid Surface

Although ^{1}H-NMR is quite useful to analyze liquid samples, the signals for solid samples are very broad and not very informative. In these cases, the NMR signals of ^{13}C, ^{31}P, and ^{29}Si should be employed. Their signals are sufficiently sharp even for solid samples and provide precise information on the structure of adducts.

Example 5.3: Complex formation on a solid surface.
When silica gel is incubated with 1-trimethoxysilylpropyl-3-guanidium chloride, guanidium residues are covalently linked onto the surface of

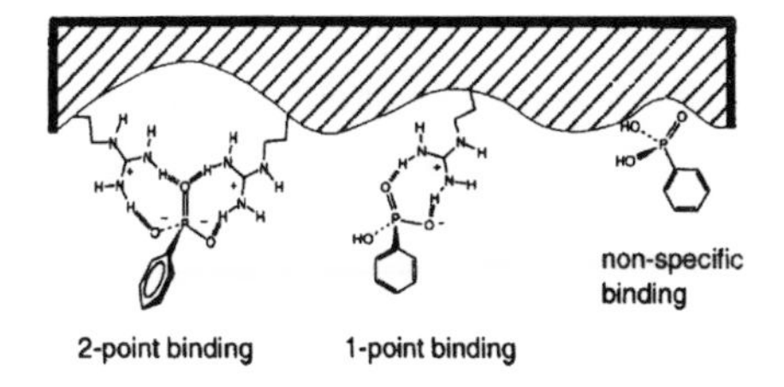

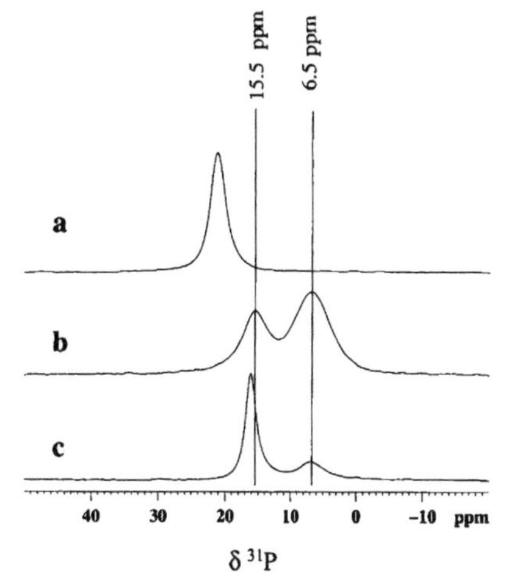

Figure 5-3 Schematic illustration of various binding modes of phenyl phosphonoic acid to guanidine-functionalized xerogel surface (upper), and solid-state ^{31}P MAS NMR (lower) of crystalline phenylphosphonoic acid (**a**) and silica xerogels imprinted with a 2:1 (**b**) and 1:1 (**c**) ratios of 1-trimethoxysilylpropyl-3-guanidium chloride. The peaks at 15.5 and 6.5 ppm are assignable to one-point and two-point bindings, respectively [5].

silica gel. Since phosphonoic acid forms a complex with guanidium residue, it can be a template for the preparation of silica gel surface, which selectively binds this guest. The mixture for this molecular imprinting has been analyzed by solid-state ^{31}P-NMR. As shown in Fig. 5.3, two peaks at 15.5 and 6.5 ppm are assigned to the P-atoms in »one-point binding« and in »two-point binding«. The solid-state ^{29}Si-NMR is consistent with the formation of these adducts.

After this reaction (covalent linkage of guanidium residues to the silica gel) is completed, the gel is treated with HCl and the phosphonoic acids are removed. In these procedures, specific binding sites for phenylphosphonoic acid are prepared on the surface of silica gel. Consistently, the binding of phenylphosphonoic acid (the guest) by this imprinted silica gel is clearly detectable by ^{31}P-NMR. Of the two kinds of binding sites, »one-point binding« is observed at around 12 ppm, and »two-point binding« at around 5 ppm. The ^{13}C-NMR with ^{13}C-enriched substrate is also useful for the analysis [5].

5.3
Determination of the Binding Constant *K* for the Formation of Monomer-Template Adduct

The values of *K*, which are important for designing effective molecular-imprinting systems, can be obtained by using these spectroscopic results. The NMR chemical shift changes are analyzed in terms of the following equation, which is straightforwardly obtained:

$$\Delta\delta_{obs} = 1/2 \cdot \Delta\delta_{max}\{(r + 1 + 1/(K\,C_0) - [(r^2 + 1 + 1/(K\,C_0)^2\,2r + 2r/(K\,C_0) + 2/(K\,C_0)]^{1/2}\} \quad (1)$$

Here, r is the monomer/template ratio and C_0 is the initial concentration of monomer (or template). The $\Delta\delta_{max}$ value is the maximal change in the chemical shift with respect to the value in free monomer (or template). By plotting $\Delta\delta_{obs}$ against r and analyzing the data by the non-linear least square method, the binding constant K and $\Delta\delta_{max}$ are determined.

It should be noted that spectroscopic titration usually shows the formation of a 1:1 adduct between template and functional monomer, even when several functional monomers interact with several parts of template. In other words, these interactions are almost independent of each other in solutions (n:1 monomer/template complexes are hardly formed therein because of the unfavorable entropy term). Even in these cases, the monomers which are interacting with the template at different positions are covalently bound to each other in the polymerization. These steps should proceed in a stepwise manner. Thus, the polymerization occurs around a site of the template. Independently, polymers are also formed at other sites of the template. Finally, these polymer domains, each of which remembers the structure of the corresponding portion of the template, are combined together in a predetermined way (dictated by the template). Accordingly, the selective binding sites can be formed from a number of functional monomers.

Example 5.4: Adduct formation between methacrylic acid (MAA) and atrazine.
On the addition of MAA in chloroform, both of the proton signals of atrazine, derived from ethylamino and isopropylamino groups, show downfield shifts (Fig. 5.4). By applying Eq. (1) to this saturation curve, the binding constant K is determined as 30 M^{-1}. Although both of the two amino groups of atrazine interact with MAA (see Fig. 6.2), only a 1:1 adduct is detected by the titration. These two interactions occur almost independently of each other, as described above. In the resultant polymers, the atrazine-binding sites would be formed from two or more carboxyl groups of MAA [6].

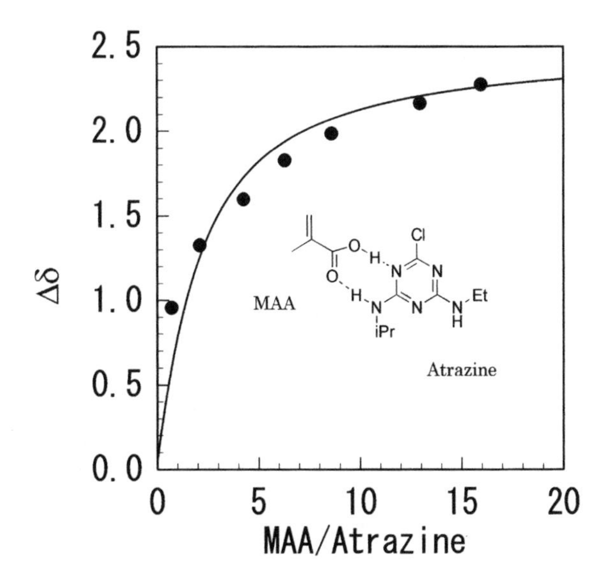

Fig. 5-4 Plots of $\Delta\delta$ against the MAA/atrazine ratio. $[\text{atrazine}]_0$ = 30 mM at 25 °C. The solid line shows the theoretical curve based on Eq. (1) (K = 30 M^{-1}) [6].

5.4
Relationship between *K* Value and Imprinting Efficiency

In order to accomplish highly efficient imprinting, the monomer-template adduct must exist in large excess with respect to free template and free functional monomer. Otherwise, non-selective polymerization would occur concurrently and diminish the efficiency of molecular imprinting. How can we realize these conditions in our imprinting experiments? By using the *K* values, we can obtain an appropriate answer to this crucial question. Table 5.1 shows the mole fractions of adduct, which are calculated under various conditions [the conditions involving sufficient amount of adducts (>70 %) are indicated by shadow]. When $K = 1\ M^{-1}$, free monomer and template exist to a significant extent even at high concentrations of both (e.g., 1 M). In general, *K* should be greater than $100\ M^{-1}$ for efficient molecular imprinting. With a *K* value

Table 5-1 Effects of *K* and concentration on the pre-organization of functional monomer

Binding constant	*Conc. of template*	*Fraction of complex (%)*[a]		
K (M^{-1})	*(mM)*	*F/T*[b] = 1	*F/T* = 5	*F/T* = 10
1	1	0.1	0.5	1.0
	10	1.0	4.7	9.0
	100	8.4	31.9	48.8
	1000	38.2	80.7[b]	90.1
10	1	1.0	4.7	9.0
	10	8.4	31.9	48.8
	100	38.2	80.7	90.1
	1000	73.0	97.6	98.9
100	1	8.4	31.9	48.8
	10	38.2	80.7	90.1
	100	73.0	97.6	98.9
	1000	90.5	99.8	99.9
1000	1	38.2	80.7	90.1
	10	73.0	97.6	98.9
	100	90.5	99.8	99.9
	1000	96.9	100	100

a Mole percent of monomer-template complex with respect to the total template. The regions involving sufficient complex formation (> 70 %) are designated by shadow.
b *F/T* = Functional monomer/template ratio.

of around 1000 M^{-1}, for example, the adducts are formed sufficiently even when the concentrations of template and monomer are only 10 mM. In most imprinted polymers, the binding sites are constructed from two or more monomer molecules. Here, the ratio of functional monomer to template (F/T) should be greater than 1. However, too large an F/T ratio should be avoided, since non-specific binding sites are formed by the functional monomer which exists in excess. As shown in Table 5.2, all the successful molecular imprinting so far reported certainly satisfies these requirements.

Table 5-2 Binding constants for template-monomer adduct formation under polymerization conditions

Template-monomer complex	*Pre-organization force*	*Solvent*	*Binding constant K M^{-1}*	*Conc. of template mM*	*Reference*
2,6-bis(acrylamido)pyridine--uracil derivatives	H-bonding	CH_2Cl_2	206	~100	*Chem. Commun.* **1995**, 2303 *Polym. Mater. Sci. Eng.*, **2000**, *82*, 69
MAA-Biotin	H-bonding	$CHCl_3$	155	65	*Anal. Chem.* **2000**, *72*, 2418
MAA-Atrazine	H-bonding	$CHCl_3$	30	66	*Anal. Chem.* **1995**, *67*, 4404
butyric acid-9-ethyladenine (Model of MAA)	H-bonding	$CHCl_3$	160	20	*J. Am. Chem. Soc.*, **1997**, *119*, 4388
β-CyD-Cholesterol	Inclusion	DMSO	550	30	*Macromolecules.* **1999**, *32*, 2265

5.5
Structure of Guest-Binding Site

In this book, it has often been stated that the sites for guest binding are formed by »freezing« pre-organized functional monomers in polymeric structures. Although this proposal has been substantiated by many experimental results, some of the readers still may be wondering whether this is really the case. Does the structure of the guest-binding site in the polymers really reflect the pre-organized structure strictly? In what manner does the template molecule affect the polymerization process and dictate the position of monomers? In order to give the answer to this question, direct observation of this »freezing« process is necessary. Mass spectroscopy is a useful tool here, since it can distinguish between polymers with different numbers of monomer units. In the following example, the template drastically accelerates the polymerization of pre-organized monomers, and promotes the formation of dimers and trimers as the guest-binding sites. In the molecular imprinting, these dimers and trimers are abundantly formed in the polymer, and these ordered assemblies selectively and efficiently bind the target guest compound.

Example 5.5: The structures of guest-binding sites in the molecularly imprinted polymer of β-cyclodextrin (β-CyD).

The polymer is prepared by crosslinking β-CyD in DMSO with toluene 2,4-diisocyanate (TDI) in the presence of cholesterol as the template. In order to obtain homogeneous samples, the amount of TDI is kept small. Matrix-assisted laser desorption/ionization time-of-flight mass spectra (MALDI-TOFMS) are presented in Fig. 5.5. In the presence of the template (a), both the dimers of β-CyD (mass number (M) = 2000–3500) and its trimers (M = 4000–4500) are efficiently formed. In its absence (b), however, virtually all the products are monomeric β-CyDs (M = 1000–2000). The template enormously accelerates the bridging between two β-CyD molecules. Each of the signals in the spectra corresponds to different amount of substitution by TDI. These analyses clearly show that dimeric β-CyDs (the binding sites for choles-

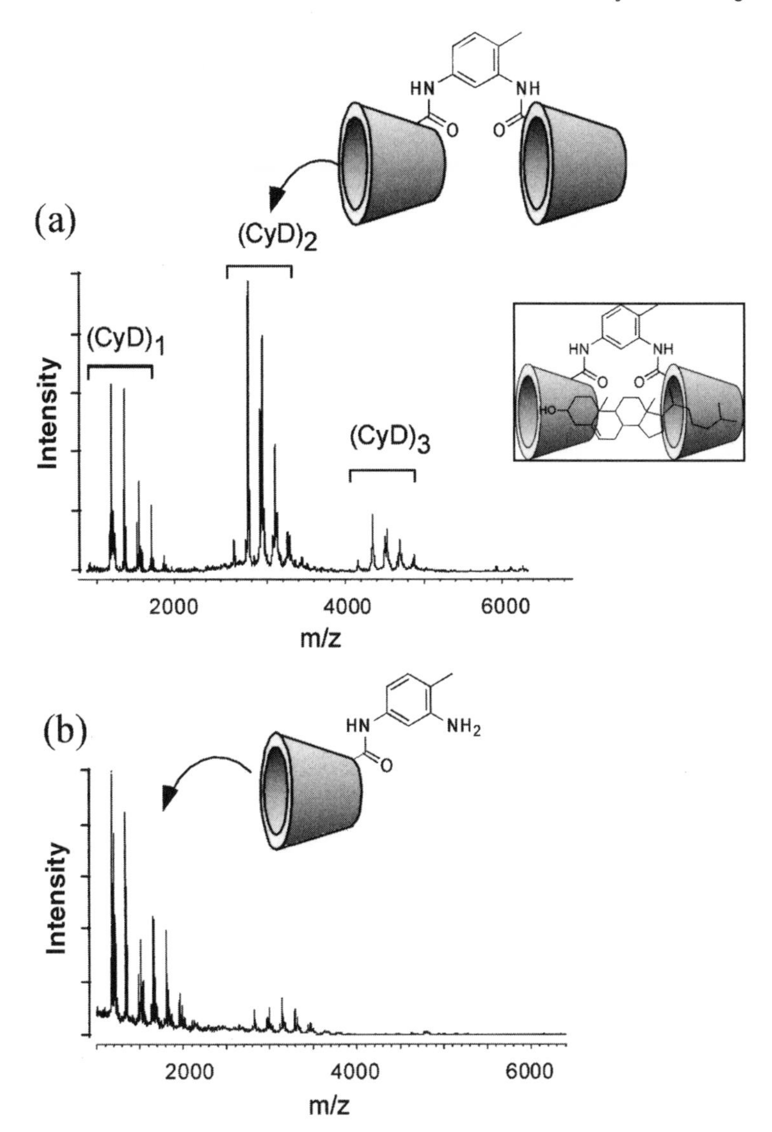

Figure 5-5 MALDI-TOFMS of β-CyD polymers crosslinked with toluene 2,4-diisocyanate in the presence (**a**) and the absence (**b**) of cholesterol. In the presence of cholesterol, dimers and trimers of β-CyD, which efficiently bind cholesterol (see the insert), are effectively formed [7].

terol) are formed by the imprinting. It should be noted that cholesterol is too large to be accommodated in the cavity of one β-CyD [7, 8].

References

1 B. Sellegren, et al., *J. Am. Chem. Soc.*, **1988**, *110*, 5853–5860.
2 H. Asanuma et al., *Supramol. Sci.*, **1998**, *5*, 417–421.
3 Takeuchi et al., *J. Chem. Soc. Chem. Commun.* **1995**, 2303–2304.
4 D. J. Duffy et al., *Polym. Mater. Sci. Eng.*, **2000**, *82*, 69–70.
5 D. Y. Sasaki et al., *Chem. Mater.* **2000**, *12*, 1400–1407.
6 J. Matsui et al., *Anal. Chem.*, **1995**, *67*, 4404–4408.
7 H. Asanuma, T. Hishiya, M. Komiyama, *Adv. Mater.*, **2000**, *12(14)*, 1019–1030.
8 H. Hishiya, H. Acanuma, M. Komiyama, *J. Am. Chem. Soc.*, **2002**, *124*, 570–575.

Chapter 6
Flow Chart of a Typical Molecular Imprinting

6.1
Introduction

In this chapter, all the processes of the molecular imprinting reaction (from the design of reaction systems to the analysis of resultant polymers) are described, so that the reader can form a clear and comprehensive mental picture of this elegant method. Non-covalent imprinting for the preparation of polymeric receptors towards atrazine (a herbicide) is taken as a typical example (Fig. 6.1).

Experimental precautions necessary to achieve the imprinting satisfactorily are also presented [1].

6.2
Choice of Agents

6.2.1
Functional Monomer

Atrazine (the template) has a triazine ring and two amino groups, all of which form hydrogen bonds with appropriate residues in aprotic solvents (Fig. 6.2). Accordingly, methacrylic acid, having a carboxylic acid (hydrogen-bonding site), is chosen as the functional monomer. Acrylic acid is also useful. The two types of interactions at the two amino

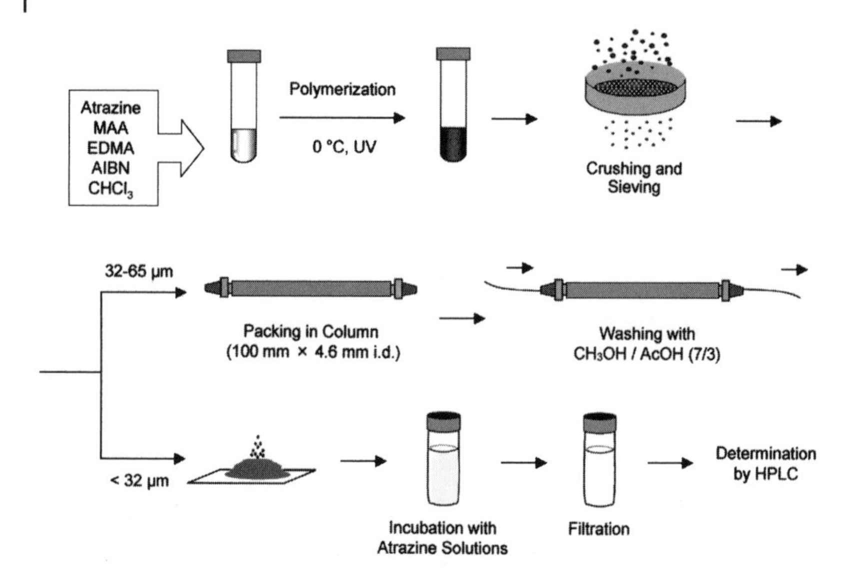

Fig. 6-1 Preparation and evaluation of atrazine-imprinted polymer

groups, shown in Fig. 6.2, occur almost independently of each other (the ^{1}H-NMR titration shows the formation of only 1:1 complex in the solutions: see Example 5.4). Because of these two interactions, atrazine and methacrylic acid form a flat and eight-membered chelate-ring which is stabilized by the resonance between a number of tautomers (Fig. 6.3). During the polymerization, the methacrylic acids showing either of these two types of interactions are bound to each other and are arranged complementarily to atrazine. As the result, high affinity and selectivity toward atrazine are achieved.

6.2.2 Polymerization Solvent

One of the most important points for successful non-covalent imprinting is to promote the formation of the non-covalent (hydrogen-bonding) adducts in the reaction mixtures. Of course, the solvent must sufficient-

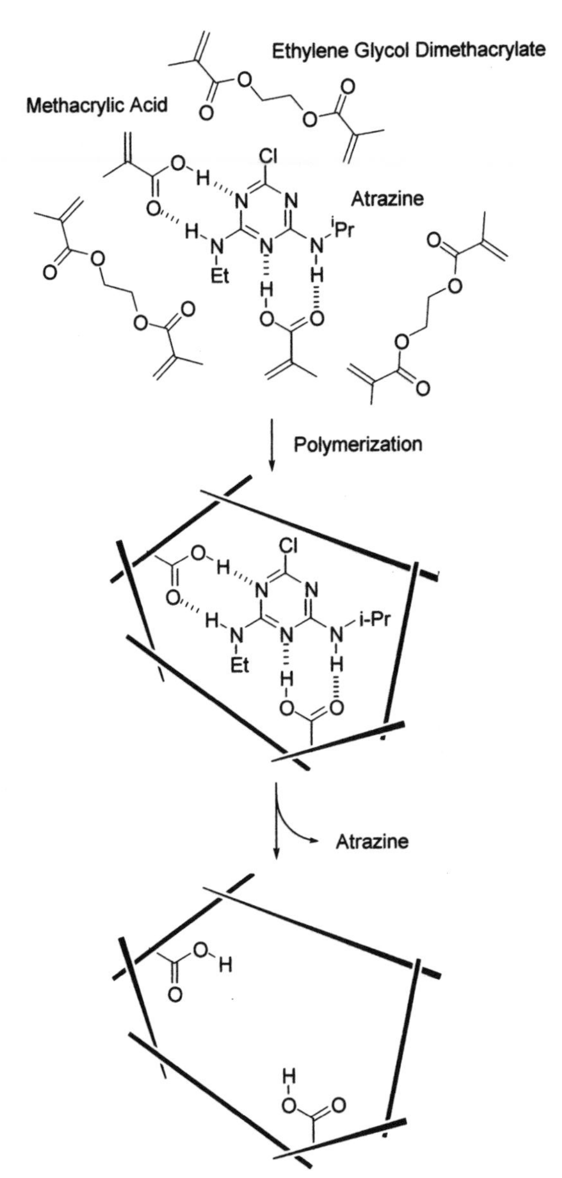

Fig. 6-2 Molecular imprinting of atrazine using methacrylic acid (MAA) as a functional monomer

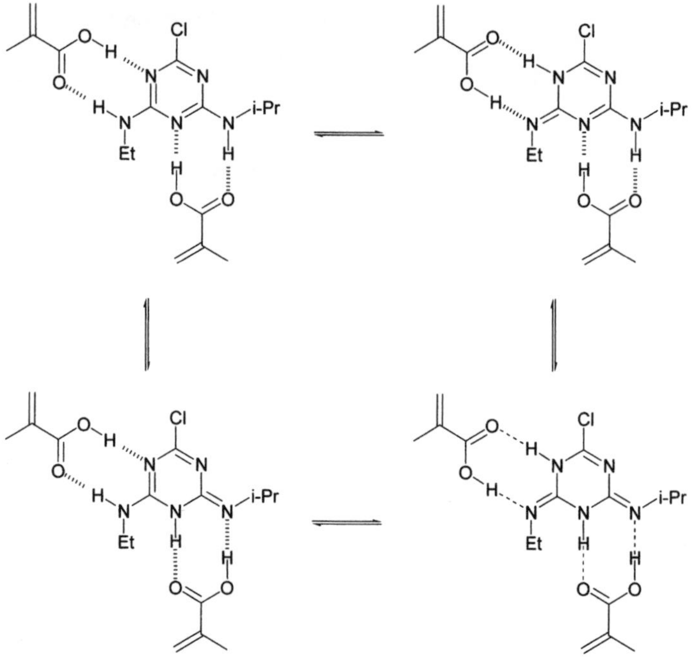

Fig. 6-3 Tautomers of hydrogen-bonding complex between atrazine and methacrylic acid

ly dissolve both the functional monomer and the template. From these viewpoints, chloroform is the best solvent. This solvent thoroughly dissolves all the reaction components and, still more significantly, does not disturb the hydrogen bonding between atrazine and methacrylic acid. Use of protic solvents such as alcohols should be avoided, since they competitively prevent the hydrogen bonding. Note that you must distill the commercially obtained chloroform before use in order to remove the ethanol, which is added as a stabilizer (see Section 3.2.3 in Chapter 3 for details).

6.2.3 Crosslinking Agent

In order to achieve effective imprinting, the functional residues (derived from the functional monomers) should be uniformly distributed in the polymer network. This situation is satisfied by choosing a crosslinking agent whose reactivity is similar to that of the functional monomer (otherwise, either the functional monomer or the crosslinking agent polymerizes predominantly with respect to the other). Here, ethylene glycol dimethacrylate (EDMA: Fig. 3.1) is chosen (note that the chemical structure of this crosslinking agent is similar to that of methacrylic acid). In the reaction mixtures, methacrylic acid and ethylene glycol dimethacrylate are copolymerized almost randomly, as desired, resulting in the required uniform distribution of the carboxylic acid groups.

6.3 Polymerization

The typical mole ratio of atrazine, methacrylic acid, and ethylene glycol dimethacrylate in chloroform is 1:3–5:25–30. Although the adduct is formed from one atrazine and two methacrylic acids (Fig. 6.2), the atrazine/methacrylic acid ratio in the mixtures is usually kept around 4 to move the equilibrium toward the formation of these hydrogen-bonding adducts. The mole ratio of crosslinking agent to the template is in general 20–30.

In order to increase the imprinting efficiency, it is desirable to achieve the polymerization at low temperatures where the non-covalent adducts between atrazine and methacrylic acid are efficiently formed. For this purpose, the initiation radicals should not be produced by thermal decomposition of radical initiator AIBN (the temperature must be 50 °C or higher here). Instead of this thermal process, AIBN (or other azo-type initiators) is decomposed by UV irradiation at 0 °C (Fig. 6.4). Under these conditions, the non-covalent adducts are abundant in the

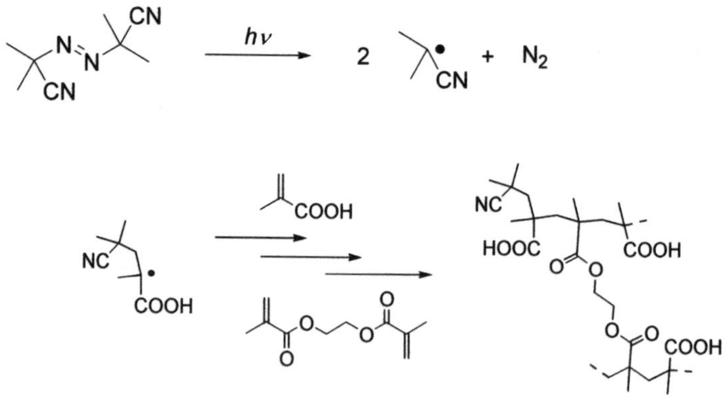

Fig. 6-4 Radical generation from AIBN by UV radiation and radical polymerization of MAA with EDMA

polymerization solutions and are predominantly polymerized. Non-selective guest-binding sites are not much formed, so that the imprinting efficiency is sufficiently high. Molecular oxygen significantly disturbs the radical polymerization. In order to remove molecular oxygen, nitrogen gas can simply be bubbled into the mixtures. Alternatively, freeze-and-thaw cycles are also useful. Both methods should give successful results.

Example 6.1: Preparation of imprinted polymer

Commercially obtained chloroform was distilled to remove ethanol (the stabilizer for storage), which would suppress hydrogen bonding if it were present. Into 25 mL of a chloroform solution of atrazine (0.36 g, 1.7 mmol), methacrylic acid (0.58 g, 6.7 mmol), ethylene glycol dimethacrylate (9.4 g, 47 mmol), and AIBN (0.12 g, 0.73 mmol) were added. Nitrogen gas was bubbled into this pre-polymerization mixture for 5 min. The polymerization tube was sealed, placed in a water bath at 0 °C, and irradiated with UV light for 4 h.

6.4
Packing of Imprinted Polymer into the HPLC Column

The polymer generated under the molecular imprinting conditions is crushed and ground in a mortar. The polymer particles are sieved, and the particles of the appropriate diameter are collected. The homogeneity of the particles is important in order to obtain stationary phases of high resolution. These particles are placed in a stainless steel column, and a solvent is passed into the column at high pressure using an HPLC pump. During this procedure, the polymer particles are densely packed in the column.

Example 6.2: Preparation of HPLC column
The polymer prepared as in Example 6.1 was ground in a mortar and dried *in vacuo*. Then the polymer particles were passed through a sieve (mesh size 32–65 μm) and suspended in acetonitrile. Too small particles were removed by decantation. The particles thus obtained were slurried and packed in a stainless-steel column-tube using an HPLC pump. In order to wash out the atrazine (used as the template for the molecular imprinting) from the column, AcOH/MeOH solution was caused to flow until a stable baseline was obtained (monitored by UV detector). The column was now available for HPLC analysis.

6.5
Quantitative Evaluation of Imprinting Efficiency

For theory, see Chapter 4.

The above-described column is used for the evaluation of imprinting efficiency (chromatographic analysis). Atrazine and an appropriate void marker are injected into the column, and the retention time for each is measured. The retention factor of atrazine is also compared with those of related compounds. The imprinting is sufficiently efficient, as confirmed by the fact that the retention factor of atrazine is selectively prolonged (see Table 6.1).

Table 6-1 Relative capacity factors for atrazine and other pesticides[a]

Guest	*Functional groups*[b]			*Relative capacity factor*	
	R_1	*R_2*	*X*	*Atrazine-Imp*	*Non-Imp*
Atrazine	C_2H_5	$CH(CH_3)_2$	Cl	1.0	0.04
Simazine	C_2H_5	C_2H_5	Cl	0.78	0.03
Cyanazine	C_2H_5	$C(CH_3)_2CN$	Cl	0.65	0.02
Prometryn	$CH(CH_3)_2$	$CH(CH_3)_2$	SCH_3	0.30	0.05
Ametryn	C_2H_5	$CH(CH_3)_2$	SCH_3	0.32	0.05

a The eluent is chloroform/acetonitrile (1/1, v/v).
b R_1, R_2, and X designate functional groups in the following structural formula:

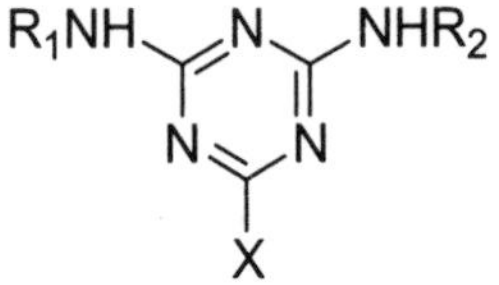

Batchwise guest-binding experiments are also used to evaluate the imprinting efficiency. The polymer particles of suitable size (usually <32 μm) are put into sample tubes, and atrazine solutions of known concentrations (typically in $CHCl_3$) are added. The resulting suspensions are stirred for a predetermined period of time. After this incubation, the polymer particles are filtered off, and the concentrations of atrazine in the solution are determined by reversed-phase HPLC. When the concentration of atrazine after the incubation is subtracted from the initial concentration, the quantity of atrazine bound by the polymer is determined. By doing similar experiments at various concentrations of atrazine, the binding isotherms are obtained and the data are analyzed by the Scatchard method.

Example 6.3: Scatchard analysis
The Scatchard plot [see Chapter 4, Eq. (4)], obtained by the methods described above, is presented in Fig. 6.5. The plot is not linear, and is composed of two straight lines (compare this non-linear plot with the linear plot in Example 4.2). This result shows that the guest-binding site

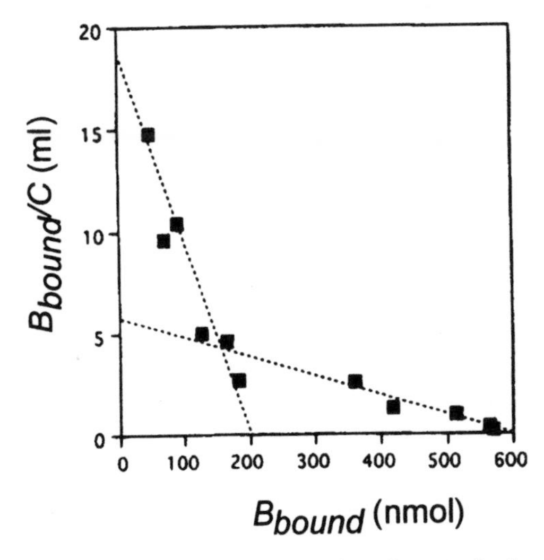

Fig. 6-5 Scatchard plot for the binding of atrazine by the atrazine-imprinted polymer [1]

in this imprinted polymer is never uniform in nature, and rather two kinds of guest-binding sites exist there. The dissociation constants K_D for these two kinds of sites are determined to be 12 μM and 100 μM, respectively. These values are 600 times and 70 times smaller (corresponding to stronger binding), respectively, than the value (7.1 mM) for the interaction between atrazine and acetic acid. Apparently, in this imprinted polymer, highly effective guest-binding sites are formed from two (or more) carboxylic acid groups which cooperatively bind the guest.

Reference

1 J. Matsui et al., *Anal. Chem.* **1995**, *67*, 4404–4408.

Chapter 7
Applications of Molecularly Imprinted Polymers

7.1
Introduction

Many compounds have now been used as template molecules in molecular imprinting. Basically, imprinted polymers can be used directly as separation media. Since all separation applications cannot be described here, some studies recently reported are listed in Table 7.1. In this chapter, only selected topics, including sensor applications, signaling polymers, molecularly imprinted sorbent assays, molecularly imprinted membranes, affinity-based solid phase extraction, *in situ* preparation of imprinted polymers, and molecularly imprinted catalysts are discussed. For the reader requiring information on other applications, there are many review articles dealing with these. Recent review articles and books are summarized in Table 7.1. For further development of molecular imprinting techniques, newly designed functional monomers would be desirable. Various functional monomers have been reported and many applications have been conducted. These are summarized in Table 7.2.

Table 7-1 Books and Reviews

Book Title
Molecular and Ionic Recognition with Imprinted Polymers, Maeda, M.; Bartsch, R. A. Eds, ACS Symp. Ser. 703, ACS, Washington, D. C., 1998.
Molecularly Imprinted Polymers: Man-Made Mimics of Antibodies and their Application in Analytical Chemistry, Sellergren, B. Ed., Elsevier, Amsterdam, 2001.

Review Title	***Reference***
Membranes in chiral separations	Kemmere, M. F.; Keurentijes, J. T. F.; Chiral Sep. Tech. (2nd Ed.) (2001), 127-150; Ed: Subramanian, G.; Publisher: Wiley-VCH Verlag GmbH, Weinheim, Germany.
Imprinted polymers with transition metal catalysts	Severin, K.; Curr. Opin. Chem. Biol. 2000, 4, 710-714.
Molecular imprinting in biological systems	Oral, E.; Peppas, N. A.; S.T.P. Pharma Sci. 2000, 10, 261-267.
Synthetic oxytocin receptors prepared by molecular imprinting	Kempe, M.; Pept. New Millennium, Proc. Am. Pept. Symp., 16th (2000) , 534-535; Eds: Fields, G. B.; Tam, J. P.; Barany, G.; Publisher: Kluwer Academic Publishers, Dordrecht, Neth.
Stationary-phase technology in separation science	Majors, R. E.; LC-GC 2000, 18, 1214, 1216, 1219-1227.
Application of imprinted synthetic polymers in binding assay development	Sellergren, B.; Andersson, L. I.; Methods (Orlando, Fla.) 2000, 22, 92-106.
Molecularly imprinted polymers and optical sensing applications	Al-Kindy, S.; Badia, R.; Suarez-Rodriguez, J. L.; Diaz-Garcia, M. E.; Critical Rev. Anal. Chem. 2000, 30, 291-309.
Tailor-made receptors by molecular imprinting	Asanuma, H.; Hishiya, T.; Komiyama, M.; Adv. Mater. (Weinheim, Ger.) 2000, 12, 1019-1030.
New configurations and applications of molecularly imprinted polymers	Bruggemann, O.; Haupt, K.; Ye, L.; Yilmaz, E.; Mosbach, K.; J. Chromatogr., A 2000, 889, 15-24.
Cyclodextrins: a versatile tool in separation science	Schneiderman, E.; Stalcup, A. M.; J. Chromatogr., B: Biomed. Sci. Appl. 2000, 745, 83-102.

Review Title	*Reference*
Molecular imprinting: developments and applications in the analytical chemistry field	Andersson, L. I.; J. Chromatogr., B: Biomed. Sci. Appl. 2000, 745, 3-13.
Recent and future developments of liquid chromatography in pesticide trace analysis	Hogendoorn, E.; Van Zoonen, P.; J. Chromatogr., A 2000, 892, 435-453.
Templated synthesis of polymers – molecularly imprinted materials for recognition and catalysis	Wulff, G.; Templated Org. Synth. (2000), 39-73. Editor: Diederich, F.; Stang, P. J.; Publisher: Wiley-VCH Verlag GmbH, Weinheim, Germany.
Alternative methods providing enhanced sensitivity and selectivity in capillary electroseparation experiments	Schweitz, L.; Petersson, M.; Johansson, T.; Nilsson, S.; J. Chromatogr., A 2000, 892, 203-217.
Imprinted polymers: versatile new tools in synthesis	Whitcombe, M. J.; Alexander, C.; Vulfson, E. N.; Synlett 2000, 911-923.
Molecularly imprinted polymers and their use in biomimetic sensors	Haupt, K.; Mosbach, K.; Chem. Rev. 2000, 100, 2495-2504.
Highly selective separations by capillary electrochromatography: molecular imprint polymer sorbents	Vallano, P. T.; Remcho, V. T.; J. Chromatogr., A 2000, 887, 125-135.
Molecular fingerprints using imprinting techniques	Dickert, F. L.; Hayden, O.; Adv. Mater. 2000, 12, 311-314.
Enantiomer separation of chiral pharmaceuticals by capillary electrochromatography	Wistuba, D.; Schurig, V.; J. Chromatogr., A 2000, 875, 255-276.
Immunochromatographic techniques – a critical review	Weller, M. G.; Fresenius' J. Anal. Chem. 2000, 366, 635-645.
Imprinted polymers with memory for small molecules, proteins, or crystals	Sellergren, B.; Angew. Chem., Int. Ed. 2000, 39, 1031-1037.
Molecular imprinting for drug bioanalysis. A review on the application of imprinted polymers to solid-phase extraction and binding assay	Andersson, L. I.; J. Chromatogr., B: Biomed. Sci. Appl. 2000, 739, 163-173.

Review Title	***Reference***
Sample handling strategies for the analysis of organic compounds in environmental water samples	Hennion, M.-C.; Tech. Instrum. Anal. Chem. 2000, 21 (Sample Handling and Trace Analysis of Pollutants), 3-71.
Chiral separation by capillary electrochromatography	Gubitz, G.; Schmid, M. G.; Enantiomer 2000, 5, 5-11.
Mid-infrared fiber optic sensors. Potential and perspectives	Mizaikoff, B.; Proc. SPIE-Int. Society Opt. English (1999), 3849(Infrared Optical Fibers and their Applications), 7-18.
Supramolecular strategies in chemical sensing	Dickert, F. L.; Sikorski, R.; Mater. Sci. English, C 1999, C10, 39-46.
Synthesis and catalysis by molecularly imprinted materials	Ramstrom, O.; Mosbach, K.; Curr. Opin. Chem. Biol. 1999, 3, 759-764.
Anion recognition: synthetic receptors for anions and their application in sensors	Snowden, T. S.; Anslyn, E. V.; Curr. Opin. Chem. Biol. 1999, 3, 740-746.
Solid-phase extraction: method development, sorbents, and coupling with liquid chromatography	Hennion, M.-C.; J. Chromatogr., A 1999, 856, 3-54.
Molecular imprinting: recent developments and the road ahead	Cormack, P. A. G.; Mosbach, K.; React. Funct. Polym. 1999, 41, 115-124.
Molecularly imprinted polymers – preparation, biomedical applications and technical challenges	Allender, C. J.; Brain, K. R.; Heard, C. M.; Prog. Med. Chem. 1999, 36, 235-291.
Preparation of beaded organic polymers and their applications in size exclusion chromatography	Lu, M. J.; Column Handb. Size Exclusion Chromatogr. (1999), 3-26. Ed: Wu, C.-S.; Publisher: Academic, San Diego, CA.
Separation and sensing based on molecular recognition using molecularly imprinted polymers	Takeuchi, T.; Haginaka, J.; J. Chromatogr., B: Biomed. Sci. Appl. 1999, 728, 1-20.
Receptor and transport properties of imprinted polymer membranes – a review	Piletsky, S. A.; Panasyuk, T. L.; Piletskaya, E. V.; Nicholls, I. A.; Ulbricht, M.; J. Membr. Sci. 1999, 157, 263-278.
Molecular recognition using surface template polymerization	Uezu, K.; Yoshida, M.; Goto, M.; Furusazki, S.; CHEMTECH 1999, 29, 12-18.

Review Title	*Reference*
Fluorescence techniques for probing molecular interactions in imprinted polymers	Wolfbeis, O. S.; Terpetschnig, E.; Piletsky, S.; Pringsheim, E.; Appl. Fluoresc. Chem., Biol. Med. (1999), 277-295. Ed: Rettig, W.; Publisher: Springer, Berlin, Germany.
Some new developments and challenges in non-covalent molecular imprinting technology	Mosbach, K.; Haupt, K.; J. Mol. Recognit. 1998, 11, 62-68.
Molecular imprinted polymers in chemical and biological sensing	Haupt, K.; Mosbach, K.; Biochem. Society Trans. 1999, 27, 344-350.
Chemical microsensors with molecularly imprinted sensitive layers	Dickert, F. L.; Greibl, W.; Sikorski, R.; Tortschanoff, M.; Weber, K.; Proc. SPIE-Int. Society Opt. English (1998), 3539(Chemical Microsensors and Applications), 114-122.
MIPs as chromatographic stationary phases for molecular recognition	Remcho, V. T.; Tan, Z. J.; Anal. Chem. 1999, 71, 248A-255A.
Towards the rational design of molecularly imprinted polymers	Nicholls, I. A.; J. Mol. Recognit. 1998, 11, 79-82.
Molecular imprinting: recent innovations in synthetic polymer receptor and enzyme mimics	Andersson, H. S.; Nicholls, I. A.; Recent Res. Dev. Pure Appl. Chem. 1997, 1, 133-157.
Plastic antibodies: developments and applications	Haupt, K.; Mosbach, K.; Trends Biotechnol. 1998, 16, 468-475.
Molecular imprints as sorbents for solid phase extraction: potential and applications	Olsen, J.; Martin, P.; Wilson, I. D.; Anal. Commun. 1998, 35, 13H-14H.
Molecular imprinted polymer for solid phase extraction of tamoxifen	Stevenson, D.; Briggs, R. J.; Hay, J.; Rashid, B.; Methodol. Surv. Bioanal. Drugs (1998), 25 (Drug Development Assay Approaches), 49-51.
Optimization of molecularly imprinted polymers for radio-ligand binding assays	Mayes, A. G.; Lowe, C. R.; Methodol. Surv. Bioanal. Drugs (1998), 25(Drug Development Assay Approaches), 28-36.
Selective SPE's using molecular imprinted polymers	Martin, P.; Wilson, I. D.; Jones, G. R.; Jones, K.; Methodol. Surv. Bioanal. Drugs (1998), 25(Drug Development Assay Approaches), 21-27.

Review Title	*Reference*
Synthesis and properties of molecular imprints of darifenacin - does molecular imprinting have a future in ultra-trace bioanalysis?	Venn, R. F.; Goody, R. J.; Methodol. Surv. Bioanal. Drugs (1998), 25(Drug Development Assay Approaches), 13-20.
New imprinting strategies	Vulfson, E. N.; Whitcombe, M. J.; Luebke, M.; Martin, L.; Esteban, I.; Ju, J.-Y.; Shin, C. S.; Klein, J.-U.; Methodol. Surv. Bioanal. Drugs (1998), 25(Drug Development Assay Approaches), 44-48.
A high-throughput screening technique employing molecularly imprinted polymers as biomimetic selectors	Bouman, M. A. E.; Allender, C. J.; Brain, K. R.; Heard, C. M.; Methodol. Surv. Bioanal. Drugs (1998), 25(Drug Development Assay Approaches), 37-43.
Molecular imprinting as an aid to drug bioanalysis	Andersson, L. I.; Methodol. Surv. Bioanal. Drugs (1998), 25(Drug Development Assay Approaches), 2-12.
Molecular imprint-based stationary phases for capillary electrochromatography	Schweitz, L.; Andersson, L. I.; Nilsson, S.; J. Chromatogr., A 1998, 817, 5-13.
Molecularly imprinted polymers for application in chemical analysis	Cormack, P. A. G.; Mosbach, K.; Am. Biotech nol. Laboratory 1998, 16, 47-48.
Applications of molecularly imprinted materials as selective adsorbents. Emphasis on enzymic equilibrium shifting and library screening	Ramstrom, O.; Ye, L.; Krook, M.; Mosbach, K.; Chromatographia 1998, 47, 465-469.
Synthesis of molecular imprinted polymer networks	Rimmer, S.; Chromatographia 1998, 47, 470-474.
Molecular imprinting in chemical sensing. Detection of aromatic and halogenated hydrocarbons as well as polar solvent vapors	Dickert, F. L.; Forth, P.; Lieberzeit, P.; Tortschanoff, M.; Fresenius' J. Anal. Chem. 1998, 360, 759-762.
Molecular imprinting technology: challenges and prospects for the future	Ramstrom, O.; Ansell, R. J.; Chirality 1998, 10, 195-209.
Selection approaches to catalytic systems	Brady, P. A.; Sanders, J. K. M.; Chem. Soc. Rev. 1997, 26, 327-336.

Review Title	*Reference*
Tailor-made materials for tailor-made applications: application of molecular imprints in chemical analysis	Ensing, K.; De Boer, T.; TrAC, Trends Anal. Chem. 1999, 18, 138-145.
Molecular imprinting for bio- and pharmaceutical analysis	Owens, P. K.; Karlsson, L.; Lutz, E. S. M.; Andersson, L. I.; TrAC, Trends Anal. Chem. 1999, 18, 146-154.
Molecular imprinted polymers for solid-phase extraction	Stevenson, D.; TrAC, Trends Anal. Chem. 1999, 18, 154-158.
Polymer- and template-related factors influencing the efficiency in molecularly imprinted solid-phase extractions	Sellergren, B.; TrAC, Trends Anal. Chem. 1999, 18, 164-174.
Validation of new solid-phase extraction materials for the selective enrichment of organic contaminants from environmental samples	Ferrer, I.; Barcelo, D.; TrAC, Trends Anal. Chem. 1999, 18, 180-192.
Molecular imprinting in chemical sensing	Dickert, F. L.; Hayden, O.; TrAC, Trends Anal. Chem. 1999, 18, 192-199.
Molecularly imprinted polymers for biosensor applications	Yano, K.; Karube, I.; TrAC, Trends Anal. Chem. 1999, 18, 199-204.
New polymeric and other types of sorbents for solid-phase extraction of polar organic micropollutants from environmental water	Masque, N.; Marce, R. M.; Borrull, F.; TrAC, Trends Anal. Chem. 1998, 17, 384-394.

Table 7-2 New Functional Monomers

Abbreviations: AA: acrylamide; MAA: methacrylic acid; TFMAA: 2-(trifluoromethyl)acrylic acid; HEMA: 2-hydroxyethyl methacrylate; MMA: methyl methacrylate; EDMA: ethylene glycol dimethacrylate; DVB: divinylbenzene; TRIM: trimethylolpropane trimethacrylate; PVA: poly(vinyl alcohol); QCM: quartz crystal microbalance; DMF: dimethylformamide; DMSO: dimethylsulfoxide; MeCN: acetonitrile; MeOH: methanol; THF: tetrahydrofuran

Template molecules	*Functional monomers, crosslinkers, porogens and other reagents*	*Remarks*	*Reference*
L-Gln	Nylon-6; phase inversion	Chiral membrane separation	[57]
D or L-Phe	2-Acryloylamido-2;2'-dimethylpropane sulfonic acid; β-cyclodextrin; *N,N'*-diacryloylpiperazine; water	Using β-cyclodextrin	[58]
D-Phe	Acrylated β-cyclodextrin; 2-acryloylamino-2-methylpropane sulfonic acid; *N,N'*-diacryloylpiperazine; water	Chiral separation in aqueous media	[59]
Z-L-Ala	Ti(O-*n*Bu)$_4$; toluene/EtOH then water	Enantioselective TiO_2 gel-based QCM sensor	[60]
Z-D-Asp	Methyl-α-D-glucopyranoside-6-acrylate; *N,N'*-methylenebisacrylamide; DMF	Chiral separation; hydrogel	[61]
Z-D-or L-Glu	Poly(hexamethylene terephthalamide/*iso*phthalamide; hexafluoro-2-propanol	Chiral membrane separation	[62]
Boc-L-Phe or (R)-(–)-2-phenylbutyric acid	*N*-(diaminoethylene)-2-methylprop-2-enamide($CHCl_3$), *N*-(3-guanidinopropyl) methacrylamide (DMF), *N*-(3-aminopropyl) methacrylamide (DMF), *N*-(2-aminoethyl) methacrylamide, 9-(β-methacryloyloxyethyl) adenine (DMF), *N*-(2-aminopropyridine) methacrylamide ($CHCl_3$) or MAA (MeCN)	Chiral separation; new functional monomers	[63]

Template molecules	*Functional monomers, crosslinkers, porogens and other reagents*	*Remarks*	*Reference*
N-Ac-L-Trp	Glu(Obz)-Phe-Phe-CH2-resin; THF; acrylonitrile; styrene	Chiral membrane separation	[64]
Boc-L-Trp	Glu(Obz)-Glu(Obz)-Glu(Obz)-CH2-resin; THF; acrylonitrile styrene	Chiral membrane separation	[65]
Boc-L-Trp or *N*-Ac-L-Trp	Glu(Obzl)-Gln-Lys(4-Cl-Z)-Leu-CH2-resin	Chiral membrane separation	[66]
D or L-Trp-OMe	*n*-Phenylphosphonic acid monododecylester; *N*-D-gluconodioleyl glutamate; DVB; toluene; phosphate buffer	Surface imprinting	[67]
D- or L-Trp-OMe	Phenylphosphonic acid monododecyl ester; L-glutamic acid dioleylester ribitol; DVB; toluene; phosphate buffer	Surface imprinting membrane	[68]
L-Trp-OMe	2-*N*-dansylethyl 3,3'-dimethylacrylate; EDMA; MeCN	Chiral fluorescence sensor; new functional monomer; competitive binding with a quencher; 4-nitrobenzaldehyde	[69]
L-Trp-OMe	Phenylphosphonic acid monododecyl ester; dioleyl-L-glutamate-D-gluconolactone amide; DVB	Chiral separation; surface molecular imprinting	[70]
N-Ac-L-Trp	Asp(OcHex)-Val-Asn-Glu(Obz)-CH_2-resin	Chiral membrane separation	[71]
N-Ac-Trp-amide or fluorescein	Bis(2-hydroxyethyl)aminopropyl-triethoxysilane; tetraethoxysilane; silica gel	Organic silane based polymer	[72]
D-Phe-D-Phe	Acryloyl-6-*O*-α-D-glucosyl-β-cyclodextrin; *N*,*N'*-methylenebisacrylamide; water	Using β-cyclodextrin	[73]

Template molecules	**Functional monomers, crosslinkers, porogens and other reagents**	**Remarks**	**Reference**
Boc-D-Ala-L-Ala *p*-nitroanilide	*N*-methacryloyl-L-valine *t*-butylamide; EDMA; $CHCl_3$	Chiral separation; new functional monomer	[74]
Albumin, IgG, lysozyme, RNase or streptavidin	Trahalose; photoresist; C_3F_6; plasma deposition to form polymeric thin films	Protein imprinting	[75]
BSA, fibrinogen, IgG, lysozyme, ribonuclease A, lactalbumin glutamine synthetase	Trahalose; photoresist; C_3F_6; plasma deposition to form polymeric thin films	Protein imprinting	[76]
(–)-cinchonidine or (+)-cinchonine	Dibenzyl-(2R;3R)-*O*-monoacryloyl tartrate; EDMA; $CHCl_3$	Chiral separation; new functional monomer	[77]
(R,S)-propranolol	MAA; TRIM; magnetic iron oxide; toluene	Chiral recognition; competitive radioligand binding assay; magnetic beads	[78]
Alloxan	2,6-Bis(acrylamido)pyridine; EDMA; $CHCl_3$+MeOH	New functional monomer	[79]
Rac-*N*-(3,5-dinitro-benzoyl)-α-methyl-benzylamine	(S)-(–)-Methacryloyl-1-naphthylethylamine; EDMA; toluene	Chiral separation; prepared with racemic template	[80]
Theophilline	Photosensitive poly(acrylonitrile-co-diethylaminodithio-carbamoylmethylstyrene	Surface imprinting membrane; photograft polymerization	[81]

Template molecules	***Functional monomers, crosslinkers, porogens and other reagents***	***Remarks***	***Reference***
Theophylline	Poly(acrylonitrile-co-acrylic acid)	Membranes prepared by phase inversion	[82]
Theophylline	MAA, TFMAA or AA; EDMA or DVB; toluene	Radioligand binding assay; investigation on molar ratios of the reagents	[83]
3-(10',12'-Penta-cosadiynamido) phenylboronic acid 4-nitrophenyl-α-D-mannopyranoside ester	10,12-Pentacosadiynoic acid; chloroform+water	Using covalently bonded templates; surface imprinting; Langmuir film	[84]
9-Ethyl adenine	5,10,15-Tris(4-isopropylphenyl)-20-(4-methacryloyloxy-phenyl)porphyrin zinc(II) complex; $CHCl_3$	New functional monomer; polymerizable metalloporphyrin	[85]
9-Ethyl adenine	5,10,15-Tris(4-isopropylphenyl)-20-(4-methacryloyloxy-phenyl)porphyrin zinc(II) complex; MAA; $CHCl_3$	New functional monomer; polymerizable metallopor-phyrin with MAA	[86]
AMP	Polyanion containing phenylboronic acid	Polyion complex based QCM sensor	[87]
CAMP	Trans-4-[*p*-(*N*;*N*-dimethylamino)styryl]-*N*-vinylbenzyl-pyridinium chloride; TRIM; HEMA; MeOH	New fluorescent functional monomer; fluorescent sensor	[88]
D- or L-threitol-[3,4-bis(bromomethyl) phenylboronic acid] 1:2 complex	[60]Fullerene; THF	Using covalently bonded templates; homogeneous nano-scale imprinting system	[89]

Template molecules	*Functional monomers, crosslinkers, porogens and other reagents*	*Remarks*	*Reference*
D-Fructose	Anthracene-boronic acid conjugate; HEMA; EDMA; MeCN	Using covalently bonded template; new functional monomer; fluorescent sensor	[90]
Glucose	Poly(o-phenylenediamine)	QCM sensor; electrosynthesized polymers	[91]
Methylated creatine	Allyl mercaptan; o-phthalaldehyde; EDMA; DMSO	New functional monomer; fluorescence assay based on bite-and-switch approach	[92]
Cholesterol	β-Cyclodextrin; hexamethylene diisocyanate or toluene 2,4-diisocyanate; DMSO	Using β-cyclodextrin	[93]
Cholesterol	Hexadecyl mercaptan	Sensor; self-assembled monolayers on Au electrode; using potassium ferricyanide	[94]
Cholesteryl(4-vinyl) phenyl carbonate	PVA; water; styrene; EDMA or DVB; AIBN; dioctyl phthalate:*n*-decane	Using covalently bonded templates; aqueous suspension polymerization	[95]
Cholesteryl(4-vinyl) phenyl carbonate or phenyl(4-vinyl)phenyl carbonate	Seed latex (MMA or MMA/EDMA or styrene or styrene/DVB); sodium lauryl sulfonate; water; peroxodisulfate	Using covalently bonded templates; core-shell emulsion polymerization	[96]
Choresterol	β-Cyclodextrin; hexamethylene diisocyanate or toluene 2,4-diisocyanate; DMSO	Using β-cyclodextrin	[97]
Choresterol	β-Cyclodextrin; toluene 2,4-diisocyanate; DMSO	Using β-cyclodextrin	[98]

Template molecules	*Functional monomers, crosslinkers, porogens and other reagents*	*Remarks*	*Reference*
Chresterol	β-Cyclodextrin; HEMA; EDMA; $CHCl_3$	Using β-cyclodextrin	[99]
Chresterol	MAA; EDMA; 3α-methacryloyldesoxycholic acid methyl ester, 3α;12α-dimethacryloyldesoxycholic acid methyl ester, 3α;7α-dimethacryloylcholic acid methyl ester or 3β-methacryloylchoresterol; MeOH or $CHCl_3$	New functional monomer	[100]
Sterol	(5-Methacryloylamino)boronophthalide; DVB; $CHCl_3$	Using covalently bonded templates; protection for regioselective modification	[101]
Atrazine (triazine herbicide)	MAA; tri(ethylene glycol)dimethacrylate; oligourethane acrylate	Conductometric sensor; membrane	[102]
Desmetryn (triazine herbicide)	2-Acrylamino-2-methylpropansulfonic acid; *N*;*N'*-methylenebis(acrylamide); porous polypropylene membranes; water	Membrane	[103]
Terbumeton (triazine herbicide)	2-Acrylamido-2-methyl-1-propane sulfonic acid, MAA, AA; *N*;*N'*-methylene-bis-acrylamide; hydrophilized polyvinylidene fluoride microfiltration membrane	SPE; membrane; surface imprinting by photo-grafting polymerization	[104]
(2,4-Diallyloxycarbonyl) benzophenone	EDMA; glycidyl methacrylate-EDMA copolymer (support polymer); benzene	Using covalently bonded templates; surface imprinting by photo-grafting polymerization	[105]
Phenyl methacrylate or *p*-cumylphenyl methacrylate	EDMA; cyclohexanol+dodecanol	Using covalently bonded templates	[106]

Template molecules	*Functional monomers, crosslinkers, porogens and other reagents*	*Remarks*	*Reference*
Phenyl methacrylate, phenyl 4-vinyl benzoate, 2-phenoxycarbonyloxy-ethyl methacrylate	EDMA; glycidyl methacrylate-EDMA copolymer (support polymer); benzene	Using covalently bonded templates; surface imprinting by photo-grafting polymerization	[107]
Phenyl methacrylate, phenyl 4-vinyl benzoate, 2-phenoxycarbonyloxy-ethyl methacrylate or (2,4-diallyloxycarbonyl) benzophenone	EDMA; glycidyl methacrylate-TRIM copolymer or glycidyl methacrylate-EDMA copolymer (support polymer); benzene	Using covalently bonded templates; surface imprinting by photo-grafting polymerization	[108]
Polycyclic aromatic hydrocarbons	Phloroglucinol, bisphenol A, *p,p'*-diisocyanatodiphenyl-methane, triisocyanate, THF	Flow system integrated with fluorescence and QCM detection	[109]
Potassium tripolyphosphate	DVB; oleylamine; oleylalcohol; HCl	Surface imprinting	[110]
Cu(II), Ni(II), Zn(II) or Co(II)	5-(4-Amino-phenylazo)-8-hydroxy quinoline; styrene acryloyl chloride copolymer; ethylene glycol; DMF	Selective extraction from aqueous solution	[111]
Zn(II)	1,12-Dodecanediol-*O,O'*-diphenylphosphonic acid; L-glutamic acid dioleylester ribitol; TRIM; toluene; 2-ethylhexyl alcohol; acetate buffer (pH 3.5); SDS; $Mg(NO_3)_2$	Surface imprinting by W/O/W emulsion polymerization for microspheres	[112]

Template molecules	*Functional monomers, crosslinkers, porogens and other reagents*	*Remarks*	*Reference*
N-(4-vinyl)benzyl triethylenetetraamine Cu(II) complex or *N,N'*-di(4-vinyl)benzyl triethylenetetra-amine Cu(II) complex	TRIM; THF	Competitive binding of Cu(II) and Zn(II) in aqueous solution	[113]
Nd(III)	Phenylphosphonic acid monododecyl ester or 2-ethylhexyl phosphonic acid mono-2-ethylhexyl ester; L-glutamic acid dioleylester ribitol; DVB; toluene+water	Surface imprinting by W/O emulsion polymerization	[114]
Cu(II)	Poly(vinylchloride-co-acrylic acid); water+THF	Thin film prepared by casting	[115]
Cu(II), Pb(II) or Li(I)	Poly(vinyl chloride-co-acrylic acid)+poly(propylene glycol); water+THF	Thin film prepared by casting	[116]
Zn(II)	1,12-Dodecanediol-*O,O'*-diphenylphosphonic acid, phenylphosphonic acid monododecyl ester or phenylphosphonic acid monohexyl ester; L-glutamic acid dioleylester ribitol; 2-ethylhexyl alcohol; DVB; toluene+water	Surface imprinting by W/O emulsion polymerization	[117]
Zn(II)	1,12-Dodecanediol-*O,O'*-diphenylphosphonic acid, 1,8-dodecanediol-*O,O'*-diphenylphosphonic acid or 1,4-dodecanediol-*O,O'*-diphenylphosphonic acid; L-glutamic acid dioleylester ribitol; 2-ethylhexyl alcohol; DVB; toluene+water	Surface imprinting by W/O emulsion polymerization	[118]
Zn(II)	1,12-Dodecanediol-*O,O'*-diphenyl phosphonic acid; L-glutamic acid dioleylester ribitol; TRIM; toluene; 2-ethylhexyl alcohol; acetate buffer (pH 3.5); SDS+$Mg(NO_3)_2$ (for W/O/W emulsion)	Surface imprinting by W/O or W/O/W emulsion polymerization for micro-spheres	[119]

Template molecules	*Functional monomers, crosslinkers, porogens and other reagents*	*Remarks*	*Reference*
Zn(II)	Dioleylphosphonic acid; L-glutamic acid dioleylester ribitol; DVB; toluene; acetate buffer (pH 3.5); SDS or PVA; $Mg(NO_3)_2$	Surface imprinting by W/O/W emulsion polymerization for microspheres	[120]
UO_2 (vinylbenzoic acid)$_2$	Styrene; divinylbenzene; pyridine	Selective extraction from aqueous solution	[121]
3-(Triethoxysilyl)propyl-benzylcarbamate, [3-(triethoxysilyl)propyl]-1,4-phenylenebis(methyl-ene)carbamate or [3-(triethoxysilyl)propyl]-1,3,5-benzenetriyltris (methylene)carbamate-benzylcarbamate	Tetraethoxysilane; EtOH; aqueous HCl	Using covalently bonded templates; sol-gel synthesis of bulk microporous silica	[122]
N-acetylated conotoxins GIII B	Silica gel; bis(2-hydroxyethyl)-aminopropyltriethoxysilane; tetraethoxysilane	Surface imprinting on silica particles	[123]
N-a-decyl-L-phenyl-alanine-2-aminopyridine	Using transition state analog; functionalized silane; tetraethoxysilane; NP-5(detergent); EtOH+cyclohexane; ammonia+ethanol	Using transition state analogue; catalytic silica particles prepared by emulsion polymerization; enantioselective hydrolysis of benzoyl-D-Arg p-nitroanilide	[124]

Template molecules	*Functional monomers, crosslinkers, porogens and other reagents*	*Remarks*	*Reference*
Phenylphosphonic acid	Tetraethoxysilane-based gel; EtOH; 1-trimethoxysilyl-propyl-3-guanidium chloride or 3-trimethoxysilylpropyl-1-trimethylammonium chloride	Surface imprinting on silica xerogels; guanidium or ammonium surface-functionalization; solid-state 31P NMR characterization	[125]
Ricin (Toxin RCA60)	Silica gel; bis(2-hydroxyethyl)aminopropyltriethoxysilane + tetraethoxysilane	Surface imprinting on silica particles	[126]
Nitrobenzene	Ni(protoporphyrin IX); CH_2Cl_2; tetrabutylammonium perchlorate	Chemically modified electrode prepared by electropolymerization	[127]
4-(4-Propyloxyphenylazo) benzoic acid	Ti(O-*n*Bu)$_4$; toluene+EtOH then water	QCM sensor, ultrathin films of titanium oxide gel prepared by surface sol-gel synthesis	[128]
Anthracene or pyrene or anthraquinone 2-sulfonic acid or pyrene 1-sulfonic acid	*p,p'*-Diisocyanatodiphenylmethane, triisocyanate, THF, bis-phenol A, 1,3,5-trihydroxybenzene	Fluorescence sensor	[129]
Cyclododecylidene pyridine-2-carboxamidrazone	Acrylamidopyridine, 5-anthracen-9-ylmethylene 3-acrylamidorhodanine, 5-(2-methoxy-naphthalen-1-ylmethylene)-3-acrylamidorhodanine or 7-hydroxy-4-methylcoumarin acrylate; TRIM or EDMA	New functional monomer; fluorescence sensor	[130]
Fresh or waste engine oil	Phloroglucinol or phloroglucinol+triethanolamine; bisphenol A; *p,p'*-diisocyanatodiphenylmethane; triisocyanate; THF	QCM sensor	[131]

Template molecules	*Functional monomers, crosslinkers, porogens and other reagents*	*Remarks*	*Reference*
N-acetyl tyrosyl-2-amino-pyridinamide or *N*-nicotin-oyl tyrosyl benzyl ester	*N*-methacryloyl L-Ser+*N*-methacryloyl L-Asp+*N*-methacryloyl L-His; Co(II)Cl_2; hydrolyzed poly(glycidyl methacrylate-EDMA) or poly(phenylmethacrylate-EDMA); EDMA; MeOH	Using substrate analo; chymotrypsin mimic; hydrolysis of *N*-acetyl tyrosyl *p*-nitrophenyl ester and *N*-benzoyl tyrosyl *p*-nitrophenyl ester; surface imprinting by photo-grafting polymerization	[132]
N-nicotinoyl tyrosyl benzyl ester	*N*-methacryloyl L-Ser, *N*-methacryloyl L-Asp and/or *N*-methacryloyl L-His; Co(II)Cl_2; EDMA; Co(II)Cl_2; poly(glycidyl methacrylate-EDMA); MeOH	Using substrate analog; chymotrypsin mimic; hydrolysis of *N*-benzyloxy carbonyl tyrosyl *p*-nitrophenyl ester; surface imprinting by photo-grafting polymerization	[133]
N-nicotinoyl tyrosyl benzyl ester	*N*-methacryloyl L-Ser, *N*-methacryloyl L-Asp, *N*-methacryloyl L-His, *N*-methacryloyl L-Cys, *N*-methacryloyl ß-Ala, 4-vinyl phenol, HEMA and/or MAA; EDMA; Co(II)Cl_2; poly(glycidyl methacrylate-EDMA); MeOH	Using substrate analog; chymotrypsin mimic; hydrolysis of *N*-benzyloxycarbo-nyl tyrosyl *p*-nitrophenyl ester; surface imprinting by photo-grafting polymerization	[134]
Bis[(1S,2S)-*N*,*N'*-di-methyl-1,2-diphenyl-ethane diamine]-1-(R)-phenylethoxy-Rh(I) complex	Diisocyanate; triisocyanate; CH_2Cl_2	Using product analog; enantioselective reduction of phenylethylketone to (R)-phenylethanol	[135]

Template molecules	*Functional monomers, crosslinkers, porogens and other reagents*	*Remarks*	*Reference*
(η^6-cymene)Ru(II) complex of *N*-(*p*-styrylsulfonyl)-ethylenediamine with diphenylphosphinate	EDMA; $CHCl_3$	Using transition state analog; benzophenone reduction by ruthenium half-sandwich complex	[136]
Cp*Rh(III)complex of (1R,2R)-*N*-(*p*-styrylsulfonyl)-1,2-diaminocyclohexane with methylphenylphosphinate	EDMA; $CHCl_3$+MeOH	Using transition state analog; Ru(III)-catalyzed asymmetric reduction of acetophenone and related aromatic ketones	[137]
Ti(IV)[(2-*tert*-butyl-6-methyl-4-vinylphenol)$_2$(NEt_2)$_2$]	Styrene; DVB; toluene	Diels-Alder reaction of 3-acryloyl-oxazolidin-2-one and cyclohexa-1,3-diene	[138]
S-ethyl, 2-(2′-imidazolyl)-4-ethenylphenyl (1-(*N-tert*-butoxycarbonylamino)-2-phenyl)ethylphosphonate, S-O-(*N-tert*-butoxycarbonyl-L-phenylalanyl)-2-(*N*-methacryloylamino)-3-(5′-imidazolyl) propanol or their derivatives	MAA; EDMA; CH_2Cl_2, $CHCl_3$ or benzene	Using substrate analog or transition state analog; enantioselective ester hydrolysis of *N-tert*-butoxycarbonyl phenylalanine *p*-nitrophenyl ester	[139]

Template molecules	*Functional monomers, crosslinkers, porogens and other reagents*	*Remarks*	*Reference*
Diphenyl phosphate	*N,N'*-diethyl(4-vinylphenyl)amidine; MMA; EDMA; MeCN, cyclohexanol+dodecanol or toluene	Using transition state analog; selective hydrolysis of diphenyl carbonate and diphenyl carbamate; bulk and suspension polymerization	[140]
N-α-t-Boc-L-His	Oleyl imidazole; L-glutamic acid dioleylester ribitol(stabilizer);DVB; toluene; Co(II)Cl_2+water	Using substrate analog; surface imprinting by W/O emulsion polymerization; hydrolysis of *N*-α-Boc-L-Ala *p*-nitrophenyl ester in isooctane/ phosphate buffer	[141]
4-(3,5-Dimethyl-phenoxyphosphonyl-methyl)-benzoic acid	*N,N'*-diethyl(4-vinylphenyl)amidine; EDMA; THF	Using transition state analog; hydrolysis of 4-(3,5-dimethyl-phenoxycarbonylmethyl)-benzoic acid to 4-carboxymethyl-benzoic acid and 3,5-dimethyl-phenol in MeCN/buffer	[142]

Table 7.3 Imprinted Polymer-Based Solid-Phase Extraction

Abbreviations: AA: acrylamide; MAA: methacrylic acid; VPy: vinyl pyridine; TFMAA: 2-(trifluoromethyl)acrylic acid; EDMA: ethylene glycol dimethacrylate; TRIM: trimethylolpropane trimethacrylate; MeCN: acetonitrile; THF: tetrahydrofuran; SPE: solid-phase extraction

			Reference
Bentazone	MAA and/or 4-VPy; EDMA; $CHCl_3$	SPE	[143]
Clenbuterol	MAA; EDMA; MeCN	SPE	[144]
Nicotine	MAA or TFMAA; EDMA or TRIM; $CHCl_3$ or MeCN	SPE	[145]
Nicotine	MAA; TRIM; CH_2Cl_2	SPE with differential pulsed elution	[146]
Pentycaine	MAA; EDMA; toluene	SPE of bupivacaine; dummy imprinting	[147]
Propranolol	MAA; EDMA or butanediol dimethacrylate; toluene	SPE	[148]
Propranolol	MAA; EDMA; toluene	SPE	[149]
Rac-propranolol	MAA; EDMA; toluene	SPE	[150]
Tamoxifen	MAA; EDMA; MeCN	SPE	[151]
Theophylline	MAA; EDMA; $CHCl_3$	SPE with pulsed elution	[152]
Theophylline	MAA; EDMA; $CHCl_3$	SPE with differential pulsed elution	[153]
Theophylline	MAA; EDMA; $CHCl_3$	SPE with differential pulsed elution	[154]
2-Amino-pyridine	MAA; $CHCl_3$; EDMA	SPE; micro-column	[155]
7-Hydroxy-coumarin	MAA; EDMA; $CHCl_3$	SPE	[156]
Indole-3-acetic acid	MAA or *N,N*-dimethyl-aminoethyl-methacrylate; EDMA; $CHCl_3$	SPE	[157]
Dibutylmel-amine (triazine herbicide analog)	MAA, EDMA; $CHCl_3$; poly(vinyl alcohol); water	SPE of atrazine; dummy imprinting; suspension polymerization	[158]

			Reference
Simazine (triazine herbicide)	MAA; EDMA; CH_2Cl_2	On-line SPE-reversed phase separation	[159]
Terbumeton (triazine herbicide)	2-Acrylamido-2-methyl-1-propane sulfonic acid; MAA; AA; *N,N'*-methylene-bis-acrylamide; hydrophilized polyvinylidene fluoride microfiltration membrane	SPE; membrane; surface imprinting by photo-grafting polymerization	[160]
Terbutylazine (triazine herbicide)	MAA; EDMA; CH_2Cl_2 or toluene	SPE	[161]
4-Nitrophenol	4-VP; EDMA; MeCN	On-line SPE coupled with RP-HPLC	[162]

7.2 Sensor Applications

7.2.1 Quartz Crystal Microbalance-based Sensors

Quartz crystal microbalance (QCM) sensors have been developed using imprinted polymers as molecular recognition elements. There are two categories: one involving the immobilization of polymer particles on the electrode of the QCM and the other *in situ* polymerization on this electrode.

Ethylene glycol dimethacrylate (EDMA)-methacrylic acid (MAA) copolymer-based imprinted polymer particles were mixed with poly(vinyl chloride) in THF, and the solution was then spread on the electrode of the QCM by spin coating. After evaporation of the THF, the polymer particles were immobilized on the surface. A phenobarbital-imprinted QCM sensor prepared in this way worked in ethanol [1], while epinephrine- and caffeine-imprinted QCMs worked in buffer solutions (pH 6.0 and pH 8.0, respectively) [2, 3].

In situ preparation of imprinted polymer films on a QCM was performed using S-propranolol as the template [4]. A pre-polymerization mixture containing MAA, trimethylolpropane trimethacrylate (TRIM, a crosslinker), the template and acetonitrile (porogen) was poured on the electrode of the QCM and immediately covered by glass and polymerized by UV irradiation. A low amount of the crosslinker (about 40 % of total monomers) was used to prepare more flexible polymer, allowing the polymer to be stably adhered on the electrode. The sensor showed enantioselective response with a selectivity factor of 5, and the detectability of S-propranolol was 50 μM in acetonitrile.

First, sialic acid- and indoleacetic acid-imprinted polymer films were prepared on the electrode of a QCM by pre-treatment with allyl mercaptan to introduce a vinyl group [5, 6]. Then, the pre-polymerization mixture containing *p*-vinylbenzeneboronic acid-sialic acid ester, EDMA, *N,N,N*-trimethylaminoethyl methacrylate, 2-hydroxyethyl methacrylate (HEMA), 2,2′-azobis(dimethylvaleronitrile), and DMF was dropped onto the electrode and covered by a trimethylchlorosilane-coated microcover glass. The polymerization was initiated by UV irradiation for 5 min. If HEMA was added, the polymer became hydrophilic, enabling it to be used not only in organic solvents but also in aqueous solution. Frequency changes were measured in air after dropping sample solution followed by washing with chloroform. The determination range was 20 to 250 nmol of sialic acid.

AMP-imprinted polyion complex layers were deposited on the electrode of a QCM [7]. At first, an anionic surface was prepared by exposing 3-mercaptopropionic acid. Then cationic polyion-containing benzeneboronic acid residues and anionic polyion-containing quaternary ammonium residues were alternatively deposited on it with a template molecule, adenosine monophosphate (AMP). When AMP was re-bound to the polyion complex, the shrinking of the imprinted polyion complex was induced, causing the frequency to change.

A carbobenzyloxy-L-alanine-titanium *n*-butoxide complex was adsorbed repeatedly to the hydroxylated surface of the electrode of the QCM by a surface sol-gel process in order to prepare ultrathin layers with molecular recognition sites [8].

Glucose-imprinted poly(*o*-phenylenediamine) was electrosynthesized on the electrode of a QCM [9]. The linear range was up to 20 mM and the response was saturated at around 100 mM. Ascorbic acid, paracetamol and cystein, which are common interferences in practical use, showed no appreciable response; however, the sensor responded slightly to fructose.

Polypyrrole film was electrosynthesized on the electrode of a QCM in the presence of L-glutamic acid as the template, followed by overoxidation, causing the film to have no charge, resulting in the removal of the template molecule [10]. L-Glutamic acid was enantioselectively bound to the film, where the ratio of bound L-glutamic acid to bound D- glutamic acid was about 10, and the binding was dependent upon the applied potential. The selectivity tests were carried out at pH 1.7 and a potential of 0.0 V vs Ag/AgCl, and it appeared that aspartic acid, phenylalanine, leucine, asparagine, cysteine, glutamine, lysine, and arginine showed almost no binding. The uptake amounts were proportional to the concentration range between 10 and 20 mM [11].

7.2.2 Electrode-type Sensors

A capacitive sensor with a molecularly imprinted polymer film as a sensitive layer has been reported. The layer was prepared by electropolymerization of phenol on a gold electrode with the template molecule, phenylalanine. The sensor capacitance was decreased by the addition of phenylalanine, but there was almost no change with glycine, tryptophan and phenol. The response time was 15 min (time for a half of the stationary value, 60 min), and the dynamic range was given as 0.5 to 8 mg/mL. The authors mentioned that temporal stability and reversibility were poor because there was no cross-linking; however, the sensor has merits such as a good selectivity and reproducibility between sensors and is suitable for a single-use sensor [12].

A herbicide, 2,4-dichlorophenoxyacetic acid (2,4-D)-imprinted polymer particles were immobilized on a disposable three-electrode system prepared by screen printing [13]. The imprinted polymer particles were suspended in methanol, and the suspension was dropped onto the working electrode. After drying, hot 2 % agarose solution was overlaid, and the electrode was covered by a thin plastic foil. After gelation, a thin agarose membrane covered the electrode. A competitive binding assay was performed on the polymer-immobilized electrode with various concentrations of 2,4-D and a competitor: 10 μM of an electroactive ligand, homogentisic acid. After incubation for 1 h in 20 mM phosphate buffer (pH 7) containing 10 % methanol, followed by briefly washing with pure water, homogentisic acid was determined by differential-pulse voltammetry. The calibration curve covered the micromolar range.

Membrane-based conductometric sensors for various target compounds have been reported [14–16]. An opposite sensor response was observed between a covalent bonding-based imprinted polymer and non-covalent bonding-based imprinted polymers. Thus, in the covalent system, the conductivity decreased with the increase of the target compound, while the non-covalent systems showed the opposite behavior. The authors suggested that the phenomenon is explained by the difference in the number of binding sites available for specific binding between the two systems, which leads to a difference in the degree of shrinking when the template was added, because in the covalent system the number of binding sites available for specific binding could be greater than that in the non-covalent systems and the binding sites could be more homogeneous. When the target molecule was bound to the polymer, the covalent bonding-based polymer had shrunk more, decreasing the size of the polymer's micropores. This may cause the decrease in electroconductivity based on ion transfer.

7.2.3 Optical Sensors

A fluorescence optical sensor for dansyl-L-phenylalanine has been reported [17]. In the optical sensor, the imprinted polymer was held in front of a fiber-optic device by a nylon net. Although the system worked well, there are some inherent problems that need to be addressed; the time required for a steady response was 4 h, which seems too long, and only fluorescent analytes could be applied to this system.

2,4-D-imprinted polymer was prepared using 4-vinylpyridine as the functional monomer and EDMA as the cross-linker on the surface of zinc selenide-attenuated total reflection elements. The on-line binding event on the chip was observed by Fourier transform infrared evanescent wave spectroscopy [18]. Although detailed calibration curves were not given, the detection range seems to be from 10^{-5} to 10^{-2} M. These results suggested that the combination of molecularly imprinted polymers with infrared evanescent wave spectroscopy could be a promising approach to developing new chemical sensors.

A surface plasmon resonance (SPR) sensor using a molecularly imprinted polymer-coated sensor chip for the detection of sialic acid was reported [19]. The thinly coated polymer was prepared by co-polymerizing *N,N,N*-trimethylaminoethyl methacrylate, HEMA, and EDMA in the presence of the *p*-vinylbenzeneboronic acid ester of sialic acid. The sensor selectively responded to a ganglioside of which sialic acid is located at the non-reducing end and gave a linear relationship from 0.1 to 1.0 mg of the ganglioside.

Theophylline-imprinted polymers prepared with MAA and EDMA were utilized for SPR sensors, in which the particles were immobilized on the silver film on the SPR sensor chip by evaporation from acetonitrile-acetic acid (99:1 v/v) containing the particles [20]. The detection limit was reported to be 0.4 mg/mL of theophylline in aqueous solution.

7.3 Signaling Polymers

A metal-complexing glucose-imprinted polymer involving ligand exchange on a triazacyclononane-copper(II) complex was reported [21]. The polymer was prepared by polymerization of the copper(II) complex of 1-(4-vinylbenzyl)-1,4,7-triazacyclononane and methyl-β-D-glucopyranoside. After removal of the methyl-β-D-glucopyranoside, the resultant polymer bound glucose selectively at alkaline pH, with the release of protons in proportion to the concentration of glucose. By operating an appropriate alkaline pH region where the buffer capacity of biological samples is small, interference with the measurement of protons released in biological samples was minimized. Equilibration of the complexation is very rapid, which suggested that this system would be suitable for continuous glucose monitoring in clinical and bioprocess applications.

A fluorescent molecularly imprinted polymer for aqueous adenosine, 3,5-cyclic monophosphate (cAMP), has been prepared [22]. Trans-4-[*p*-(*N,N*-dimethylamino)styryl]-*N*-vinylbenzylpyridinium chloride was used as a functional monomer for interacting with the template molecules by electrostatic and aryl stacking. In addition to the functional monomer, a large amount of HEMA was admixed in order to increase the hydrophilicity of the polymer and thus supplement the hydrogen bond formation with the template, as it has already been reported that the use of HEMA improved the recognition ability of sialic acid-imprinted polymer in aqueous solution [23]. When cAMP was bound to the polymer, the fluorescence of the polymer was quenched in proportion to the amount of cAMP added. In contrast, adding cAMP did not change the fluorescence of a control polymer prepared without the template cAMP. For a structurally similar molecule, guanosine 3,5-cyclic monophosphate (cGMP), the fluorescence quenching was not observed, and it appears that cAMP is selectively bound to the imprinted polymer and the binding event is readable by the degree of quenching. It should be noted that the association constant of the free functional monomer with cAMP is reported to be about 14 M^{-1}, whereas that of the imprinted polymer is 3.5×10^5 M^{-1}, suggesting that the three-dimensional polymer network affects the specific binding and enhances the affinity.

More recently, a diastereoselective molecularly imprinted fluorescent polymer for (–)-cinchonidine was prepared by the combined use of methacrylic acid and vinyl-substituted zinc(II) porphyrin monomer as functional monomers [24]. Compared to the reference imprinted polymers using either MAA or zinc(II) porphyrin as a functional monomer, the imprinted polymer prepared with both MAA and the porphyrin monomer showed higher binding ability for (–)-cinchonidine in chromatographic tests. Scatchard analysis gave a higher association constant (1.14 10^7 M^{-1}) than these reference polymers.

The zinc(II) porphyrin-based imprinted polymer showed fluorescence quenching according to the binding of (–)-cinchonidine, and the quenching was significant in the low concentration range, suggesting that the high-affinity binding sites contain the porphyrin residue. A plot of the relative fluorescence intensity against the log of (–)-cinchonidine concentrations showed a linear relationship. The zinc(II) porphyrin-based polymer, which appeared to act as a fluorescence sensor, selectively responded by binding of the template molecule.

A nucleobase derivative, 9-ethyladenine-imprinted polymer capable of spectroscopic change based upon molecular recognition has also been prepared using the zinc(II) porphyrin-based functional monomer [25].

7.4 Molecularly Imprinted Sorbent Assays

Molecularly imprinted sorbent assays represent one of the most typical applications of biomimetic use, where imprinted polymers are used as substitutes of natural antibodies in immunoassays. The assays usually involve competitive binding of an analyte with a certain quantity of labeled ligands, in which the labeled ligand unbound is proportional to the analyte added. Because dissociation constants of common imprinted polymers are around 10^{-6}–10^{-9} M, competitive binding assays could easily be performed. In practice, many molecularly imprinted sorbent assays have been developed for biologically active compounds, including theophylline, diazepam [26], S-propranolol [27], morphine, Leu-enkephalin [28], cyclosporin A [29], yohimbine [30], methyl-α-glucoside [31], corticosteroid [32], atrazine [33, 34], and 2,4-D [35].

This technique may be useful as a supplement to antibody-based assays, because imprinted polymers can be stable under severe conditions where natural molecules cannot survive, such as in organic solvents or acidic/basic solutions, at high temperatures and so on. Currently, molecularly imprinted sorbent assays are usually performed with radio-ligands. Although non-isotopic assays have been desired, these would not be achieved easily, because binding sites in imprinted polymers are fitted to the template, and labeled templates are not usually suitable for the binding sites. Efforts toward the development of non-isotopic molecularly imprinted sorbent assays have been made, and several studies have been reported [36–38]. However, further work should be aimed at the development of more reliable methods for practical use.

A fluorescence detection system for sialic acid has also been reported [39]. For the detection of sialic acid, *o*-phthalaldehyde reagent was used for the fluorescence measurement of amino residues in the polymer matrix. According to the authors, when sialic acid was bound to the polymer, the fluorescence intensity was increased, because the binding of the template increased the permeability of the *o*-phthalaldehyde reagent because of the swelling change. The increase in fluorescence was proportional to the amount of sialic acid bound.

Optical detection of an antibiotic, chloramphenicol, based on competitive displacement of a chloramphenicol-methyl red conjugate bound to a chloramphenicol-imprinted polymer with free chloramphenicol has been demonstrated [40]. A flow injection system in conjunction with a 10 cm stainless steel column packed with the imprinted polymer and acetonitrile as a carrier solution containing chloramphenicol-methyl red conjugate was constructed. The dye conjugate released by displacement by free chloramphenicol was monitored at 460 nm. The signals were proportional to the concentration of free chloramphenicol injected, and the calibration range of this system included the therapeutic range of a chloramphenicol. This concept of flow displacement systems could be applicable not only for chloramphenicol determination but also for other template molecules.

7.5 Molecularly Imprinted Membranes

A nucleotide base-imprinted polymer membrane has been reported in which methacrylic acid was used as a functional monomer for the imprinting of an adenine derivative, 9-ethyladenine [41]. A free-standing film was prepared by polymerizing a DMF solution containing methacrylic acid and ethylene glycol dimethacrylate on a silanized glass slide at 65–70 °C under nitrogen atmosphere.

A series of enantioselective imprinted polymer membranes for amino acid and peptide derivatives were prepared using oligopeptides as functional monomers [42–45]. A tetrahydrofuran solution containing a template molecule, a functional monomer of a peptide derivative attached on polystyrene resin that is commonly used in solid-phase peptide synthesis, copolymer of acrylonitrile and styrene, was poured into a flat laboratory dish and left for 24 h to remove the solvent.

Theophylline-imprinted membranes were prepared by phase inversion of poly(acrylonitrile-co-acrylic acid). The copolymer and the template molecule, theophylline, dissolved in dimethylsulfoxide, was spread on a glass plate about 0.1 mm thick and coagulated in water. After removal of the theophylline by washing with 0.1 %(v/v) acetic acid, a theophylline-imprinted polymer membrane was obtained [46]. Instead of poly(acrylonitrile-co-acrylic acid), Nylon-6 was also used in this technique to prepare L-glutamine-imprinted polymer membranes [47].

An ultrathin-film composite membrane selective for theophylline has been reported [48]. The theophylline-imprinted polymer was prepared inside pores of a microporous alumina support membrane with a thickness of 500 nm and a pore size of 20 nm, in which pores of the membrane were filled by the polymerization solution containing the template theophylline, methacrylic acid, and ethylene glycol dimethacrylate, and the membrane was illuminated with UV light for 1 h, followed by immersion in methanol containing 10 %(v/v) acetic acid to remove the template and any excess monomer. Because the membrane is extremely thin, the flux rate is high, being at least two orders of magnitude higher than the flux rate in the 9-ethyladenine-imprinted membrane described above [41].

Surface modifications of porous membranes with molecularly imprinted polymers have been reported. The membranes were photografted with a functional monomer, 2-acrylamido-2-methylpropanesulfonic acid and a cross-linker, *N,N*-methylenebis(acrylamide) in the presence of a template, desmetryn, in water [49].

7.6 Affinity-based Solid-phase Extraction

Molecularly imprinted solid-phase extraction (MI-SPE) of triazine herbicides has been reported using an atrazine-imprinted polymer [50]. This procedure consists of three steps: (1) sample loading where the polymer works as a reversed phase system because the aqueous sample is loaded on the column, (2) washing with dichloromethane, when the system is changed to a hydrogen-bonding-based affinity mode in which triazine herbicides can be selectively retained in the polymer while other structurally unrelated impurities are washed off, (3) recovery of triazine herbicides with methanol, when hydrogen bonding is significantly weakened because of the interference in the hydrogen bond formation by the methanol. By employing such MI-SPE, simazine (0.1 ppm), one of the herbicides commonly used in golf courses in Japan, was selectively concentrated approximately 60-fold with over 90 % recovery from a mixture of simazine, asulam, mecoprop, propyzamide and iprodione (0.1 ppm each, 500 mL of aqueous solution).

Although MI-SPE showed a good performance, an inevitable and crucial problem exists with respect to its practical use. After the polymer synthesis, atrazine is removed from the polymer to generate MAA-based binding sites complementary to atrazine. The problem is that atrazine cannot be removed completely and may leak out of the polymer during the SPE operation.

In order to avoid the problem, a technique using dummy template molecules was developed for the preparation of triazine herbicide-selective polymers [51], whereby alkylmelamines instead of atrazine are used as the template species for the polymer synthesis. These are capable of intermolecular interaction with methacrylic acid in a similar fashion, and have no influence on the analysis of atrazine even though they cannot be eliminated from the polymer before use.

Many imprinted polymers have been applied to MI-SPE, and recent developments are listed in Table 7.3.

7.7
In Situ Preparation of Imprinted Polymers

In situ molecular imprinting is a convenient way to prepare imprinted polymers. Here, imprinted polymers are prepared in a place where the polymers are subsequently utilized. In general, molecularly imprinted polymers are prepared by bulk polymerization, and block polymers obtained are broken to pieces, ground, sieved and packed in a column. These experimental procedures are extremely tedious and time-consuming. The procedure also results in polymer particles of irregular size and shape, which may have a negative influence on column efficiency.

In order to overcome these problems, the first *in situ* molecular imprinting was performed to prepare a molecularly imprinted chromatographic stationary phase [52]. A column filled with all the reagents necessary for molecular imprinting was heated in a water bath, resulting in a ready-to-use column filled with a continuous imprinted polymer rod, through which the eluent flows because of the porosity of imprinted polymers. An enantioselective polymer rod was obtained by *in situ* molecular imprinting using L-phenylalanine anilide as the template, and exhibited a separation factor of 1.7.

Continuous polymer rods, however, usually exhibit higher back pressure than conventional stationary phases. Therefore, a choice among polymerization solvents known to behave as pore formers has to be carefully made so that the resultant polymers are porous enough to give good flow properties of the eluent. To date, the most successful pore formers in imprinted rod preparation are reported to be cyclohexanol-dodecanol, though these solvents may be less favorable for hydrogen bonding and ionic interaction between a template and functional monomers.

Another *in situ* preparation of molecularly imprinted columns employs dispersion polymerization, whereby agglomerated polymer particles are obtained [53]. The procedure is similar to the rod preparation; a mixture of the chemicals for the polymer preparation, such as a template, a functional monomer, a crosslinker, a porogen and an initiator is placed in a column and heated to cause polymerization. This method also requires polar solvents, such as cyclohexanol-dodecanol and isopropanol-water, to obtain aggregated polymer particles of well-defined micron sizes. A crucial difference lies in the volume of the porogen used, this being larger in dispersion polymerization than in rod preparation.

The *in situ* molecular imprinting protocol employing dispersion polymerization can be advantageous. The dispersion polymer can be removed from a column and re-packed if a column is damaged after repeated use. Back pressure of agglomerated polymer particles is less problematical, and therefore this *in situ* method can be used for a wider range of analytical techniques.

In order to synthesize the imprinted polymer that gives the best imprint effect for a target molecule, it is necessary to synthesize a large number of polymers by changing various reaction conditions. The three main time-consuming and complicated operations are the preparation of the pre-polymerization mixture, the sieving and washing of the imprinted polymers generated, and the batch rebinding tests for evaluation. These procedures may be simplified by synthesizing thin film polymers at the bottom of glass vials using a programmed liquid handler. This combinatorial molecular imprinting is a method that applies the concept of combinatorial chemistry to molecularly imprinting. It is possible at the same time to make libraries of imprinted polymers in which amounts of template, functional monomer, and crosslinking reagents can be changed using an automatic liquid handler. By programming the amount of each reagent to be added, the time-consuming operations can be performed automatically. UV or thermal polymerization can be carried out in the glass vials as they are. The polymers are automatically washed by dispensing and aspirating a washing solvent, and a sample solution is dispensed. After incubation, supernatants are analyzed by HPLC. Using these labor-saving operations, it is possible to synthesize and evaluate a large number of polymers simultaneously and automatically, and the polymerization conditions for enhancing the imprint effect can be optimized conveniently.

Using the above-described automated system, several functional monomers have been screened for the development of molecularly imprinted polymers for the herbicides atrazine and ametryn [54]. According to the results, MAA appears to be more effective for developing the affinity in the atrazine-imprinted polymers. In contrast, 2-(trifluoromethyl)acrylic acid is more effective for ametryn imprinting.

Catalytic polymers for atrazine decomposition were also screened [55]. Among the monomers tested, 2-sulfoethyl methacrylate was found to catalyze the conversion of atrazine into a low-toxicity compound atraton, and these results significantly contributed to the following detailed experimental plan.

7.8 Molecularly Imprinted Catalysts

One of the most attractive applications would be molecularly imprinted catalysts. In principle, such catalysts could be prepared if substrate, product or transition-state analogs could be used as template molecules, since to natural catalytic antibodies are produced in a similar way. Since molecularly imprinted polymers are considered to be analogous to antibodies in that binding sites are tailor-made, catalytic antibody-like activity in imprinted polymers could also be conceived, enabling an »artificial catalytic antibody« with the advantageous features of synthetic molecules to be produced.

The first »artificial catalytic antibody« for the hydrolysis of p-nitrophenyl acetate was prepared using a transition state analog, *p*-nitrophenyl methylphosphonate as the template [56]. Polyvinylimidazole cross-linked by 1,4-dibromobutane in the presence of the transition state analog exhibited 1.7-fold higher activity for catalyzing the planned hydrolysis reaction than that of a non-imprinted reference polymer. The activity was inhibited by the addition of the template, suggesting that imprinted cavities successfully operated as catalytic sites.

Since then, many attempts have been made to produce molecularly imprinted catalysts. However, because of the inherent strong affinity to template molecules, these almost always showed lower turnover than that of natural enzymes. A major breakthrough would be necessary for further development of molecularly imprinted catalysts. Recent investigations into this subject are summarized in Table 7.4.

Table 7-4 Molecularly Imprinted Catalysts

Abbreviations: MAA: methacrylic acid; VPy: vinyl pyridine; MMA: methyl methacrylate; EDMA: ethylene glycol dimethacrylate; DVB: divinylbenzene; MeCN: acetonitrile; MeOH: methanol; THF: tetrahydrofuran

Template molecules	***Functional monomers, crosslinkers, porogens and other reagents***	***Remarks***	***Reference***
N-α-Decyl-L-phenylalanine-2-aminopyridine	Using transition state analog; functionalized silane; tetraethoxysilane; NP-5(detergent); EtOH+cyclohexane; ammonia+ethanol	Using transition state analog; catalytic silica particles prepared by emulsion polymerization; enantio-selective hydrolysis of benzoyl-D-Arg *p*-nitroanilide	[163]
p-Nitrophenyl phosphate	Vinyl imidazole; DVB; MeCN	Using transition state analog; hydrolysis of *p*-nitrophenyl acetate	[164]
N-acetyl tyrosyl-2-amino-pyridinamide or *N*-nicotinoyl tyrosyl benzyl ester	*N*-methacryloyl L-Ser+*N*-methacryloyl L-Asp+*N*-methacryloyl L-His; Co(II)Cl2; hydrolyzed poly(glycidyl methacrylate-EDMA) or poly(phenylmethacrylate-EDMA); EDMA; MeOH	Using substrate analog; chymotrypsin mimic; hydrolysis of *N*-acetyl tyrosyl *p*-nitrophenyl ester and *N*-benzoyl tyrosyl *p*-nitrophenyl ester; surface imprinting by photo-grafting polymerization	[165]
N-nicotinoyl tyrosyl benzyl ester	*N*-Methacryloyl L-Ser, *N*-methacryloyl L-Asp and/or *N*-methacryloyl L-His; Co(II)Cl_2; EDMA; Co(II)Cl_2; poly(glycidyl methacrylate-EDMA); MeOH	Using substrate analog; chymotrypsin mimic; hydrolysis of *N*-benzyloxycarbonyl tyrosyl *p*-nitrophenyl ester; surface imprinting by photo-grafting polymerization	[166]

Template molecules	*Functional monomers, crosslinkers, porogens and other reagents*	*Remarks*	*Reference*
N-nicotinoyl tyrosyl benzyl ester	*N*-methacryloyl L-Ser, *N*-methacryloyl L-Asp, *N*-methacryloyl L-His, *N*-methacryloyl L-Cys, *N*-methacryloyl ß-Ala, 4-vinyl phenol, HEMA and/or MAA; EDMA; Co(II)Cl_2; poly(glycidyl methacrylate-EDMA); MeOH	Using substrate analog; chymotrypsin mimic; hydrolysis of N-benzyloxy-carbonyl tyrosyl p-nitrophenyl ester; surface imprinting by photo-grafting polymerization	[167]
Indole or 2-amino-5,6-dimethylbenzimidazole	4-VPy; EDMA; CH_2Cl_2	Using substrate analog; benzisoxazole isomerization to hydroxybenzonitrile in EtOH/water	[168]
Bis[(1S,2S)-*N*,*N'*-dimethyl-1,2-diphenylethane diamine]-1-(R)-phenylethoxy-Rh(I) complex	Diisocyanate; triisocyanate; CH_2Cl_2	Using product analog; enantioselective reduction of phenylethylketone to (R)-phenylethanol	[169]
(η^6-cymene)Ru(II)complex of *N*-(*p*-styrylsulfonyl)-ethylenediamine with diphenylphosphinate	EDMA; $CHCl_3$	Using transition state analog; benzophenone reduction by ruthenium half-sandwich complex	[170]
Cp*Rh(III)complex of (1R,2R)-*N*-(*p*-styrylsulfonyl)-1,2-diaminocyclohexane with methylphenylphosphinate	EDMA; $CHCl_3$+MeOH	Using transition state analog; Ru(III)catalyzed asymmetric reduction of acetophenone and related aromatic ketones	[171]
Ti(IV)[(2-*tert*-butyl-6-methyl-4-vinyl-phenol)$_2$(NEt_2)$_2$]	Styrene; DVB; toluene	Diels-Alder reaction of 3-acryloyl-oxazolidin-2-one and cyclohexa-1,3-diene	[172]

Template molecules	*Functional monomers, crosslinkers, porogens and other reagents*	*Remarks*	*Reference*
S-ethyl, 2-(2'-imidazolyl)-4-ethenylphenyl (1-(*N-tert*-butoxycarbonylamino)-2-phenyl)ethylphosphonate, S-O-(*N-tert*-butoxycarbonyl-L-phenylalanyl)-2-(*N*-methacryloylamino)-3-(5'-imidazolyl) propanol or their derivatives	MAA; EDMA; CH_2Cl_2, $CHCl_3$ or benzene	Using substrate analog or transition state analog; enantioselective ester hydorolysis of *N*-tert-butoxycarbonyl phenylalanine *p*-nitrophenyl ester	[173]
Diphenyl phosphate	*N,N'*-diethyl(4-vinylphenyl)amidine; MMA; EDMA; MeCN, cyclohexanol+dodecanol or toluene	Using transition state analog; selective hydrolysis of diphenyl carbonate and diphenyl carbamate; bulk and suspension polymerization	[174]
N-α-t-Boc-L-His	Oleyl imidazole; L-glutamic acid dioleylester ribitol (stabilizer); DVB; toluene; Co(II)Cl_2+water	Using substrate analog; surface imprinting by W/O emulsion polymerization; hydrolysis of *N*-α-Boc-L-Ala *p*-nitrophenyl ester in isooctane/phosphate buffer	[175]
4-(3,5-Dimethyl-phenoxy-phosphonylmethyl)-benzoic acid	*N,N'*-diethyl(4-vinylphenyl)amidine; EDMA; THF	Using transition state analog; hydrolysis of 4-(3,5-dimethyl-phenoxycarbonylmethyl)-benzoic acid to 4-carboxymethyl-benzoic acid and 3,5-dimethyl-phenol in MeCN/buffer	[176]
Atrazine	MAA; 2-sulfoethylaminomethacrylate; EDMA; $CHCl_3$	Combinatorial molecular imprinting for atrazine-decomposing polymers	[177]

References

1 Peng, H., Liang, C., He, D., Nie, L., Yao, S., Anal. Lett. 2000, 33, 793.
2 Liang, C., Peng, H., Zhou, A., Nie, L., Yao, S., Anal. Chim. Acta 2000, 415, 135.
3 Liang, C., Peng, H., Bao, X., Nie, L., Yao, S., Analyst 1999, 124, 1781.
4 Haupt, K., Noworyta, K., Kutner, W., Anal. Commun. 1999, 36, 391.
5 Kugimiya, A., Yoneyama, H., Takeuchi, T., Electroanalysis 2000, 12, 1322.
6 Kugimiya, A., Takeuchi, T., Electroanalysis 1999, 11, 1158.
7 Kanekiyo, Y., Inoue, K., Ono, Y., Sano, M., Shinkai, S., Reinhoudt, D. N., J. Chem. Soc., Perkin Trans. 2 1999, 2719.
8 Lee, S.-W., Ichinose, I., Kunitake, T., Chem. Lett. 1998, 1193.
9 Malitesta, C., Losito, I., Zambonin, P. G., Anal. Chem. 1999, 71, 1366.
10 Deore, B., Chen, Z., Nagaoka, T., Anal. Chem. 2000, 72, 3989.
11 Deore, B., Chen, Z., Nagaoka, T., Anal. Sci. 1999, 15, 827.
12 Panasyuk, T. L., Mirsky, V. M., Piletsky, S. A., Wolfbeis, O. S., Anal. Chem. 1999, 71, 4609.
13 Kroeger, S., Turner, A. P. F., Mosbach, K., Haupt, K., Anal. Chem. 1999, 71, 3698.
14 Piletsky, S. A., Piletskaya, E. V., Panasyuk, T. L., El'skaya, A. V., Levi, R., Karube, I., Wulff, G., Macromolecules 1998, 31, 2137.
15 Sergeyeva, T. A., Piletsky, S. A., Panasyuk, T. L., El'skaya, A. V., Brovko, A. A., Slinchenko, E. A., Sergeeva, L. M., Analyst 1999, 124, 331.
16 Sergeyeva, T. A., Piletsky, S. A., Brovko, A. A., Slinchenko, E. A., Sergeeva, L. M., El'skaya, A. V., Anal. Chim. Acta 1999, 392, 105.
17 Kriz, D., Ramström, O., Svensson, A., Mosbach, K. Anal. Chem. 1995, 67, 2142.
18 Jakusch, M., Janotta, M., Mizaikoff, B., Mosbach, K., Haupt, K., Anal. Chem. 1999, 71, 4786.
19 Kugimiya, A., Takeuchi, T., Biosens. Bioelectron. 2001, 16, 1059.
20 Lai, E. P. C., Fafara, A., Vandernoot, V. A., Kono, M., Polsky, B., Can. J. Chem. 1998, 76, 265.
21 Chen G., Guan Z, Chen C.-T., Fu L., Sundaresan V., Arnold F. H., Nat. Biotechnol., 15 (1997) 354.
22 Turkewitsch, P., Wandelt, B., Darling, G. D., Powell, W. S., Anal. Chem. 1998, 70, 2025.
23 Kugimiya, A., Takeuchi, T., Matsui, J., Ikebukuro, K., Yano, K., Karube, I. Anal. Lett. 1996, 29, 1099.
24 Takeuchi, T., Mukawa, T., Matsui, J., Higashi, M., Shimizu, K. D., Anal. Chem. 2001, 73, 3869.
25 Matsui, J., Higashi, M., Takeuchi, T., J. Am. Chem. Soc. 2000, 122, 5218.
26 Vlatakis G., Andersson L. I., Müller R., Mosbach K., Nature (London) 1993, 361, 645.
27 (a) Andersson L. I., Anal. Chem. 1996, 68, 111. (b) Bengtsson H., Roos U., Andersson L. I., Anal. Commun. 1997, 34, 233.
28 Andersson L. I., Müller R., Vlatakis G., Mosbach K., Proc. Natl. Acad. Sci. U. S. A. 1995, 92, 4788.
29 Senholdt M., Siemann M., Mosbach K., Andersson L. I., Anal. Lett. 1997, 30, 1809.

30 Berglund J., Nicholls I. A., Lindbladh C., Mosbach K., Bioorg. Med. Chem. Lett. 1996, 6, 2237.
31 Mayes A. G., Andersson L. I., Mosbach K., Anal. Biochem. 1994, 222, 483.
32 Ramström O., Ye L., Mosbach K., Chem. Biol. 1996, 3, 471.
33 Muldoon M. T., Stanker L. H., J. Agric. Food Chem. 1995, 43, 1424.
34 Siemann M., Andersson L. I., Mosbach K., J. Agric. Food Chem. 1996, 44, 141.
35 Haupt K., Dzgoev A., Mosbach K., Anal. Chem. 1998, 70, 628.
36 Kriz D., Mosbach K., Anal. Chim. Acta 1995, 300, 71.
37 Levi R., McNiven S., Piletsky S. A., Cheong S.-H., Yano K., Karube I., Anal. Chem. 1997, 69, 2017.
38 Piletsky S. A., Piletskaya E. V., El'skaya A. V., Levi R., Yano K., Karube I., Anal. Lett. 1997, 30, 445.
39 Piletsky S. A., Piletskaya E. V., Yano K., Kugimiya A., Elgersma A. V., Levi R., Kahlow U., Takeuchi T., Karube I., Panasyuk T. I., El'skaya A. V., Anal. Lett. 1996, 29, 157.
40 McNiven, S., Kato, M., Levi, R., Yano, K., Karube, I., Anal. Chim. Acta 1998, 365, 69.
41 Mathew-Krotz J., Shea K. J., J. Am. Chem. Soc. 1996, 118, 8154.
42 Yoshikawa M., Izumi J., Kitao T., Sakamoto S., Macromolecules 1996, 29, 8197.
43 Yoshikawa M., Izumi J., Kitao T., Chem. Lett. 1996, 611.
44 Yoshikawa M., Izumi J., Kitao T., Sakamoto S., Makromol. Rapid Commun. 1997, 18, 761.
45 Yoshikawa M., Fujisawa T., Izumi J., Kitao T., Sakamoto S., Anal. Chim. Acta 1998, 365, 59.
46 Wang, H. Y., Kobayashi, T., Fujii, N., J. Chem. Technol. Biotechnol. 1997, 70, 355.
47 Reddy, P. S., Kobayashi, T., Fujii, N., Chem. Lett. 1999, 293.
48 Hong, J.-M., Anderson, P. E., Qian, J., Martin, C. R., Chem. Mater. 1998, 10, 1029–1033.
49 Piletsky, S. A., Matuschewski, H., Schedler, U., Wilpert, A., Piletska, E. V., Thiele, T. A., Ulbricht, M., Macromolecules 2000, 33, 3092.
50 Matsui, J., Okada, M., Tsuruoka, M., Takeuchi, T., Anal. Commun. 1997, 34, 85.
51 Matsui, J., Fujiwara, K., Takeuchi, T., Anal. Chem. 2000, 72, 1810–1813. Matsui, J., Fujiwara, K., Ugata, S., Takeuchi, T., J. Chromatogr. A 2000, 889, 25.
52 Matsui, J., Kato, T. Takeuchi, T., Suzuki, M., Yokoyama, K., Tamiya, E., Karube, I., Anal. Chem. 1993, 65, 2223.
53 Sellergren, B., J. Chromatogr. A 1994, 673, 133.
54 Takeuchi, T., Fukuma, D., Matsui, J., Anal. Chem. 1999, 71, 285.
55 Takeuchi, T., Fukuma, D., Matsui, J., Mukawa, T., Chem. Lett. 2001, 530.
56 Robinson D. K., Mosbach K., J. Chem. Soc., Chem. Commun. 1989, 969.
57 Reddy, P. S.; Kobayashi, T.; Fujii, N., Chem. Lett. 1999, 293–294.
58 Piletsky, S. A.; Andersson, H. S.; Nicholls, I. A., Macromolecules 1999, 32, 633–636.
59 Piletsky, S. A.; Andersson, H. S.; Nicholls, I. A., J. Mol. Recognit. 1998, 11, 94–97.
60 Lee, S.-W.; Ichinose, I.; Kunitake, T., Chem. Lett. 1998, 1193–1194.

61 Liu, X.-C.; Dordick, J. S., J. Polym. Sci., Part A: Polym. Chem. 1999, 37, 1665–1671.

62 Kondo, Y.; Yoshikawa, M.; Okushita, H., Polym. Bulletin (Berlin) 2000, 44, 517–524.

63 Spivak, D.; Shea, K. J., J. Org. Chem. 1999, 64, 4627–4634.

64 Yoshikawa, M.; Fujisawa, T.; Izumi, J., Macromol. Chem. Phys. 1999, 200, 1458–1465.

65 Yoshikawa, M.; Ooi, T.; Izumi, J.-I., Eur. Polym. J. 2000, Volume Date 2001, 37, 335–342.

66 Yoshikawa, M.; Fujisawa, T.; Izumi, J.-I.; Kitao, T.; Sakamoto, S., Anal. Chim. Acta 1998, 365, 59–67.

67 Yoshida, M.; Hatate, Y.; Uezu, K.; Goto, M.; Furusaki, S., Colloids Surf., A 2000, 169, 259–269.

68 Yoshida, M.; Uezu, K.; Goto, M.; Furusaki, S., J. Appl. Polym. Sci. 2000, 78, 695–703.

69 Liao, Y.; Wang, W.; Wang, B., Bioorg. Chem. 1999, 27, 463–476.

70 Yoshida, M.; Uezu, K.; Goto, M.; Furusaki, S.; Takagi, M., Chem. Lett. 1998, 925–926.

71 Yoshikawa, M.; Fujisawa, T.; Izumi, J.-I.; Kitao, T.; Sakamoto, S., Sen'i Gakkaishi 1998, 54, 77–84.

72 Lulka, M. F.; Chambers, J. P.; Valdes, E. R.; Thompson, R. G.; Valdes, J. J., Anal. Lett. 1997, 30, 2301–2313.

73 Asanuma, H.; Kajiya, K.; Hishiya, T.; Komiyama, M., Chem. Lett. 1999, 665–666.

74 Yano, K.; Nakagiri, T.; Takeuchi, T.; Matsui, J.; Ikebukuro, K.; Karube, I., Anal. Chim. Acta 1997, 357, 91–98.

75 Shi, H.; Tsai, W.-B.; Garrison, M. D.; Ferrari, S.; Ratner, B. D., Nature 1999, 398, 593–597.

76 Shi, H.; Ratner, B. D., J. Biomed. Mater. Res. 2000, 49, 1–11.

77 Knutsson, M.; Andersson, H. S.; Nicholls, I. A., J. Mol. Recognit. 1998, 11, 87–90.

78 Ansell, R. J.; Mosbach, K., Analyst 1998, 123, 1611–1616.

79 Yano, K.; Tanabe, K. Takeuchi, T.; Matsui, J.; Ikebukuro, K.; Karube, I., Anal. Chim. Acta 1998, 363, 111–117.

80 Hosoya, K.; Shirasu, Y.; Kimata, K.; Tanaka, N., Anal. Chem. 1998, 70, 943–945.

81 Wang, H. Y.; Kobayashi, T.; Fujii, N., J. Chem. Technol. Biotechnol. 1997, 70, 355–362.

82 Kobayashi, T.; Wang, H. Y.; Fujii, N., Anal. Chim. Acta 1998, 365, 81–88.

83 Yilmaz, E.; Mosbach, K.; Haupt, K., Anal. Commun. 1999, 36, 167–170.

84 Miyahara, T.; Kurihara, K., Chem. Lett. 2000, 1356–1357.

85 Matsui, J.; Tachibana, Y.; Takeuchi, T., Anal. Commun. 1998, 35, 225–227.

86 Matsui, J.; Higashi, M.; Takeuchi, T., J. Am. Chem. Soc. 2000, 122, 5218–5219.

87 Kanekiyo, Y.; Inoue, K.; Ono, Y.; Sano, M.; Shinkai, S.; Reinhoudt, D. N., J. Chem. Soc., Perkin Trans. 2 1999, 2719–2722.

88 Turkewitsch, P.; Wandelt, B.; Darling, G. D.; Powell, W. S., Anal. Chem. 1998, 70, 2025–2030.

89 Ishi-I, T.; Iguchi, R.; Shinkai, S., Tetrahedron 1999, 55, 3883–3892.

90 Wang, W.; Gao, S.; Wang, B., Org. Lett. 1999, 1, 1209–1212.

91 Malitesta, C.; Losito, I.; Zambonin, P. G., Anal. Chem. 1999, 71, 1366–1370.

92 Subrahmanyam, S.; Piletsky, S. A.; Piletska, E. V.; Chen, B.; Day, R.;

Turner, A. P. F., Adv. Mater. (Weinheim, Ger.) 2000, 12, 722–724.
93 Asanuma, H.; Kakazu, M.; Shibata, M.; Hishiya, T.; Komiyama, M., Chem. Commun. 1997, 1971–1972.
94 Piletsky, S. A.; Piletskaya, E. V.; Sergeyeva, T. A.; Panasyuk, T. L.; El'skaya, A. V., Sens. Actuators, B 1999, B60, 216–220.
95 Flores, A.; Cunliffe, D.; Whitcombe, M. J.; Vulfson, E. N., J. Appl. Polym. Sci. 2000, 77, 1841–1850.
96 Perez, N.; Whitcombe, M. J.; Vulfson, E. N., J. Appl. Polym. Sci. 2000, 77, 1851–1859.
97 Asanuma, H.; Kakazu, M.; Shibata, M.; Hishiya, T. Komiyama, M., Supramol. Sci. 1998, 5, 417–421.
98 Hishiya, T.; Shibata, M.; Kakazu, M.; Asanuma, H.; Komiyama, M., Macromolecules 1999, 32, 2265–2269.
99 Sreenivasan, K., J. Appl. Polym. Sci. 1998, 70, 15–18.
100 Sellergren, B.; Wieschemeyer, J.; Boos, K.-S.; Seidel, D., Chem. Mater. 1998, 10, 4037–4046.
101 Alexander, C.; Smith, C. R.; Whitcombe, M. J.; Vulfson, E. N., J. Am. Chem. Soc. 1999, 121, 6640–6651.
102 Sergeyeva, T. A.; Piletsky, S. A.; Panasyuk, T. L.; El'skaya, A. V.; Brovko, A. A.; Slinchenko, E. A.; Sergeeva, L. M., Analyst 1999, 124, 331–334.
103 Piletsky, S. A.; Matuschewski, H.; Schedler, U.; Wilpert, A.; Piletska, E. V.; Thiele, T. A.; Ulbricht, M., Macromolecules 2000, 33, 3092–3098.
104 Sergeyeva, T. A.; Matuschewski, H.; Piletsky, S. A.; Bendig, J.; Schedler, U.; Ulbricht, M., J. Chromatogr., A 2001, 907, 89–99.
105 Joshi, V. P.; Kulkarni, M. G.; Mashelkar, R. A., J. Chromatogr., A 1999, 849, 319–330.
106 Joshi, V. P.; Karmalkar, R. N.; Kulkarni, M. G.; Mashelkar, R. A., Ind. Eng.. Chem. Res. 1999, 38, 4417–4423.
107 Joshi, V. P.; Karode, S. K.; Kulkarni, M. G.; Mashelkar, R. A., Chem. Eng.. Sci. 1998, 53, 2271–2284.
108 Joshi, V. P.; Kulkarni, M. G.; Mashelkar, R. A., Chem. Eng. Sci. 2000, 55, 1509–1522.
109 Dickert, F. L.; Tortschanoff, M.; Bulst, W. E.; Fischerauer, G. Anal. Chem. 1999, 71, 4559–4563.
110 Fujiwara, I.; Maeda, M.; Takagi, M., Anal. Sci. 2000, 16, 407–412.
111 Ray, A.; Gupta, S. N., J. Appl. Polym. Sci. 1998, 67, 1215–1219.
112 Yoshida, M.; Hatate, Y.; Uezu, K.; Goto, M.; Furusaki, S., J. Polym. Sci., Part A: Polym. Chem. 2000, 38, 689–696.
113 Singh, A.; Puranik, D.; Guo, Y.; Chang, E. L., React. Funct. Polym. 2000, 44, 79–89.
114 Araki, K.; Yoshida, M.; Uezu, K.; Goto, M.; Furusaki, S., J. Chem. Eng.. Japan 2000, 33, 665–668.
115 Kanekiyo, Y.; Sano, M.; Ono, Y.; Inoue, K.; Shinkai, S., J. Chem. Soc., Perkin Trans. 2 1998, 2005–2008.
116 Kanekiyo, Y.; Inoue, K.; Ono, Y.; Shinkai, S., Tetrahedron Lett. 1998, 39, 7721–7724.
117 Yoshida, M.; Uezu, K.; Goto, M.; Furusaki, S., Macromolecules 1999, 32, 1237–1243.
118 Yoshida, M.; Uezu, K.; Nakashio, F.; Goto, M., J. Polym. Sci., Part A: Polym. Chem. 1998, 36, 2727–2734.
119 Yoshida, M.; Uezu, K.; Goto, M.; Furusaki, S., J. Appl. Polym. Sci. 1999, 73, 1223–1230.

120 Uezu, K.; Nakamura, H.; Goto, M.; Nakashio, F.; Furusaki, S., J. Chem. Eng. Japan 1999, 32, 262–267.

121 Bae, S. Y.; Southard, G. L.; Murray, G. M., Anal. Chim. Acta 1999, 397, 173–181.

122 Katz, A.; Davis, M. E., Nature (London) 2000, 403, 286–289.

123 Iqbal, S. S.; Lulka, M. F.; Chambers, J. P.; Thompson, R. G.; Valdes, J. J., Mater. Sci. English, C 2000, C7, 77–81.

124 Markowitz, M. A.; Kust, P. R.; Deng, G.; Schoen, P. E.; Dordick, J. S.; Clark, D. S.; Gaber, B. P., Langmuir 2000, 16, 1759–1765.

125 Sasaki, D. Y.; Alam, T. M., Chem. Mater. 2000, 12, 1400–1407.

126 Lulka, M. F.; Iqbal, S. S.; Chambers, J. P.; Valdes, E. R.; Thompson, R. G.; Goode, M. T.; Valdes, J. J., Mater. Sci. English, C 2000, C11, 101–105.

127 Panasyuk, T.; Dall'Orto, V. C.; Marrazza, G.; El'skaya, A.; Piletsky, S.; Rezzano, I.; Mascini, M., Anal. Lett. 1998, 31, 1809–1824.

128 Lee, S.-W.; Ichinose, I.; Kunitake, T., Langmuir 1998, 14, 2857–2863.

129 Dickert, F. L.; Besenboeck, H.; Tortschanoff, M., Adv. Mater. (Weinheim, Ger.) 1998, 10, 149–151.

130 Rathbone, D. L.; Su, D.; Wang, Y.; Billington, D. C., Tetrahedron Lett. 2000, 41, 123–126.

131 Dickert, F. L.; Forth, P.; Lieberzeit, P. A.; Voigt, G., Fresenius' J. Anal. Chem. 2000, 366, 802–806.

132 Lele, B. S.; Kulkarni, M. G.; Mashelkar, R. A., Polymer 1999, 40, 4063–4070.

133 Lele, B. S.; Kulkarni, M. G.; Mashelkar, R. A., React. Funct. Polym. 1999, 39, 37–52.

134 Lele, B. S.; Kulkarni, M. G.; Mashelkar, R. A., React. Funct. Polym. 1999, 40, 215–229.

135 Locatelli, F.; Gamez, P.; Lemaire, M., J. Mol. Catal. A: Chem. 1998, 135, 89–98.

136 Polborn, K.; Severin, K., Chem. Commun. 1999, 2481–2482.

137 Polborn, K.; Severin, K., Eur. J. Inorg. Chem. 2000, 1687–1692.

138 Santora, B. P.; Larsen, A. O.; Gagne, M. R., Organometallics 1998, 17, 3138–3140.

139 Sellergren, B.; Karmalkar, R. N.; Shea, K. J., J. Org. Chem. 2000, 65, 4009–4027.

140 Strikovsky, A. G.; Kasper, D. Gruen, M.; Green, B. S.; Hradil, J.; Wulff, G., J. Am. Chem. Soc. 2000, 122, 6295–6296.

141 Toorisaka, E.; Yoshida, M.; Uezu, K.; Goto, M.; Furusaki, S., Chem. Lett. 1999, 387–388.

142 Wulff, G.; Gross, T.; Schonfeld, R., Angew. Chem., Int. Ed. Engl. 1997, 36, 1962–1964.

143 Baggiani, C.; Giraudi, G.; Giovannoli, C.; Vanni, A.; Trotta, F., Anal. Commun. 1999, 36, 263–266.

144 Berggren, C.; Bayoudh, S.; Sherrington, D.; Ensing, K., J. Chromatogr., A 2000, 889, 105–110.

145 Zander, A.; Findlay, P.; Renner, T.; Sellergren, B.; Swietlow, A., Anal. Chem. 1998, 70, 3304–3314.

146 Mullett, W. M.; Lai, E. P. C.; Sellergren, B., Anal. Commun. 1999, 36, 217–220.

147 Andersson, L. I., Analyst 2000, 125, 1515–1517.

148 Olsen, J.; Martin, P.; Wilson, I. D.; Jones, G. R., Analyst 1999, 124, 467–471.

149 Martin, P.; Wilson, I. D.; Jones, G. R., J. Chromatogr., A 2000, 889, 143–147.
150 Martin, P.; Wilson, I. D.; Jones, G. R., Chromatographia 2000, 52, S19–S23.
151 Rashid, B. A.; Briggs, R. J.; Hay, J. N.; Stevenson, D., Anal. Commun. 1997, 34, 303–305.
152 Mullett, W. M.; Lai, E. P. C., Anal. Chem. 1998, 70, 3636–3641.
153 Mullett, W. M.; Lai, E. P. C., Microchem. J. 1999, 61, 143–155.
154 Mullett, W. M.; Lai, E. P. C., J. Pharm. Biomed. Anal. 1999, 21, 835–843.
155 Mullett, W. M.; Dirie, M. F.; Lai, E. P. C.; Guo, H.; He, X., Anal. Chim. Acta 2000, 414, 123–131.
156 Walshe, M.; Howarth, J.; Kelly, M. T.; O'Kennedy, R.; Smyth, M. R., J. Pharm. Biomed. Anal. 1997, 16, 319–325.
157 Kugimiya, A.; Takeuchi, T., Anal. Chim. Acta 1999, 395, 251–255.
158 Matsui, J.; Fujiwara, K.; Ugata, S.; Takeuchi, T., J. Chromatogr., A 2000, 889, 25–31.
159 Bjarnason, B.; Chimuka, L.; Ramstroem, O., Anal. Chem. 1999, 71, 2152–2156.
160 Sergeyeva, T. A.; Matuschewski, H.; Piletsky, S. A.; Bendig, J.; Schedler, U.; Ulbricht, M., J. Chromatogr., A 2001, 907, 89–99.
161 Ferrer, I.; Lanza, F.; Tolokan, A.; Horvath, V.; Sellergren, B.; Horvai, G.; Barcelo, D., Anal. Chem. 2000, 72, 3934–3941.
162 Masque, N.; Marce, R. M.; Borrull, F.; Cormack, P. A. G.; Sherrington, D. C., Anal. Chem. 2000, 72, 4122–4126.
163 Markowitz, M. A.; Kust, P. R.; Deng, G.; Schoen, P. E.; Dordick, J. S.; Clark, D. S.; Gaber, B. P., Langmuir 2000, 16, 1759–1765.
164 Kawanami, Y.; Yunoki, T.; Nakamura, A.; Fujii, K.; Umano, K.; Yamauchi, H.; Masuda, K., J. Mol. Catal. A: Chem. 1999, 145, 107–110.
165 Lele, B. S.; Kulkarni, M. G.; Mashelkar, R. A., Polymer 1999, 40, 4063–4070.
166 Lele, B. S.; Kulkarni, M. G.; Mashelkar, R. A., React. Funct. Polym. 1999, 39, 37–52.
167 Lele, B. S.; Kulkarni, M. G.; Mashelkar, R. A., React. Funct. Polym. 1999, 40, 215–229.
168 Liu, X.-C.; Mosbach, K., Macromol. Rapid Commun. 1998, 19, 671–674.
169 Locatelli, F.; Gamez, P.; Lemaire, M., J. Mol. Catal. A: Chem. 1998, 135, 89–98.
170 Polborn, K.; Severin, K., Chem. Commun. 1999, 2481–2482.
171 Polborn, K.; Severin, K., Eur. J. Inorg. Chem. 2000, 1687–1692.
172 Santora, B. P.; Larsen, A. O.; Gagne, M. R., Organometallics 1998, 17, 3138–3140.
173 Sellergren, B.; Karmalkar, R. N.; Shea, K. J., J. Org. Chem. 2000, 65, 4009–4027.
174 Strikovsky, A. G.; Kasper, D. Gruen, M.; Green, B. S.; Hradil, J.; Wulff, G., J. Am. Chem. Soc. 2000, 122, 6295–6296.
175 Toorisaka, E.; Yoshida, M.; Uezu, K.; Goto, M.; Furusaki, S., Chem. Lett. 1999, 387–388.
176 Wulff, G.; Gross, T.; Schonfeld, R., Angew. Chem., Int. Ed. Engl. 1997, 36, 1962–1964.
177 Takeuchi, T.; Fukuma, D.; Matsui, J.; Mukawa, T., Chem. Lett. 2001, 530–531

Chapter 8
Recent Challenges and Progress

8.1
Introduction

Since »molecular imprinting« is a comprehensive concept and easy to carry out, numerous strategies for still wider and more versatile applications have been proposed and examined. Although fundamental concepts have already been realized and several polymers are now proceeding to practical application (see Chapter 7), the molecular imprinting technique is still growing. As the targets are being extended from angstrom-sized simple molecules to nanometer-sized complicated ones, new strategies and concepts are being developed. In this chapter, typical examples of recent challenges for stepping up this technology are introduced.

8.2
Molecular Imprinting in Water

Molecular imprinting polymers are often compared to natural antibodies. However, there still exist wide gaps between them. One of these is the solvent in which receptors bind the targets: most of the imprinted polymers reported so far function only in organic solvents, while natural antibodies do so in water. Recent chemistry is being directed to nonorganic and water-based system (so-called green chemistry). Nevertheless, reports on imprinted polymers for aqueous systems are still limited. The main barriers which prevent imprinting in water are as follows:

1. Hydrogen bonds, which are preferentially used for pre-organization of templates and functional monomers, are easily destroyed in bulk water because of competition with this solvent [1].
2. Water-soluble conventional crosslinking agents (e.g., *N,N'*-methylenebisacrylamide) cannot sufficiently reinforce the polymer, and therefore the obtained polymers are not stiff enough for the stationary phase of HPLC.

The first barrier is associated with the nature of hydrogen bonds. Water itself is both a hydrogen donor and a hydrogen acceptor, and forms hydrogen bonds with hosts or guests. Since solvent molecules exist in large amounts, they competitively destroy hydrogen-bonded host-guest adducts. Here, the reader might wonder why natural antibodies, enzymes, and DNA can recognize the target molecules precisely through hydrogen-bond formation *even in water*. The answer lies in the hydrophobic spheres provided by these natural polymers for the formation of hydrogen bonds. In our molecular imprinting, functional monomer and template cannot be precisely preorganized in water unless this problem is somehow solved. The second barrier is crucial for practical applications. Chemical modification of crosslinking agent is not very successful: introduction of an aromatic moiety certainly promotes the mechanical strength, but at the same time decreases the solubility in water. In order to overcome these barriers, new methodologies are proposed. The followings are some recent examples.

8.2.1
Hydrophobic Interactions for the Preorganization of Functional Monomers and Template in Water

In contrast with hydrogen bonding, hydrophobic interactions work in aqueous solutions. By choosing appropriate functional monomers, we can utilize these interactions for the preorganization in water. Cyclodextrins (CyDs) (cyclic oligomers of glucose units) are promising candidates for this purpose: they have apolar cavities (5–8 Å in interior diameter and 7 Å in depth) and form inclusion complexes with apolar guests in bulk water [2]. By placing multiple CyDs complementarily to the hydrophobic groups of the target molecule, the organized assembly as a whole can bind the target selectively in water.

Example 8.1: Preparation of imprinted CyD polymers (Fig. 8.1) *[3]*

The vinyl monomer of CyD is synthesized by the ester exchange reaction of *m*-nitrophenyl acrylate with β-CyD or α-CyD in water. The imprinted polymers are prepared in water by a conventional radical co-polymerization of the vinyl monomer of CyD with *N,N'*-methylenebisacrylamide (MBAA) as crosslinker in the presence of various templates: acryloyl CyD (300 μmol) and template (150 μmol) are dissolved in 15 mL of Tris buffer solution ([Tris] = 5 mM, pH 8.0). After stirring for a few minutes, the polymerization is started by adding MBAA (3.0 mmol) and potassium persulfate (35 mg) under nitrogen at 50 °C. The system becomes opaque as the polymerization proceeds. After stirring for 2 h, the obtained white precipitate is collected and

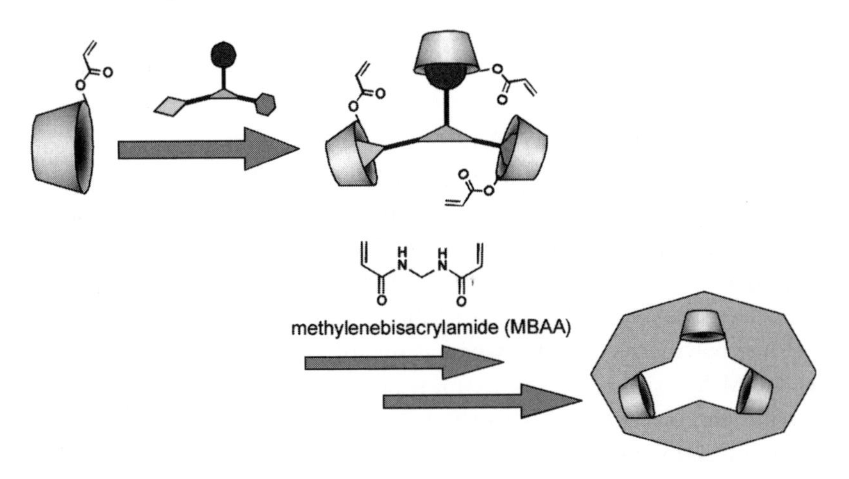

Fig. 8-1 Molecular imprinting of CyD monomer in the presence of a template in water

Table 8-1 Binding constants (K) and maximal amounts of binding (A) determined from Scatchard plots

Template (Guest)	*Imprinted β-CyD*			*Non-imprinted*
	$K/10^2\ M^{-1}$	$A/\mu mol\ g^{-1}$	$K/10^2\ M^{-1}$	$A/\mu mol\ g^{-1}$
Vancomycin	6.3	44	2.4	52
Cefazolin	3.2	52	1.4	68
D-Phe-D-Phe	70	24	12	50

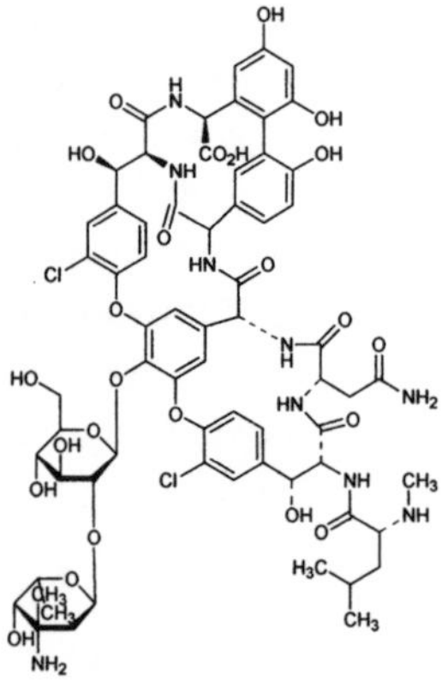

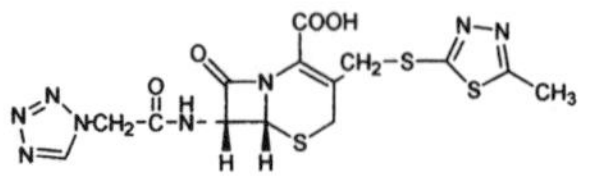

Cefazolin

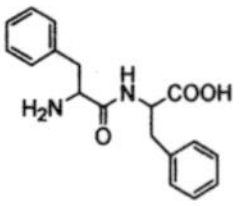

Vancomycin

D-Phe-D-Phe or L-Phe-L-Phe

washed with a large amount of hot water and acetone. By this treatment, the template molecule is completely removed from the polymer.

The imprinting of CyD in water efficiently enhances the binding activity toward the template molecule. The binding constant (K) and the maximal amount of adsorption (A) are determined from the Scatchard plots (see Chapter 4). As listed in Table 8.1, the K of the vancomycin-imprinted polymer is 630 M^{-1}, whereas that of the non-imprinted polymer is 240 M^{-1}. A similar increase in K is attained upon imprinting with cefazolin and D-Phe-D-Phe. This promotion of binding is attributed to the immobilization of two or more CyDs in a complementary manner to the hydrophobic groups of the template, so that the organized assembly of CyDs as a whole binds the target. Note that all the templates used here have more than two hydrophobic groups. The imprinting effect is also demonstrated by the conventional HPLC analysis in the following example, which is one of the answers for the second barrier described above.

Example 8.2: Imprinting of CyD on the surface of silica gel support (Fig. 8.2.) *[4]*
Although a conventional crosslinking agent such as MBAA does not provide the polymer with enough stiffness as a stationary phase of HPLC, a silica gel support can reinforce this soft polymer. By introducing vinyl group on the silica-gel surface, the imprinted polymer can easily be immobilized on its surface by the conventional polymerization.

Commercially available silica gel (e.g., Nucleosil 300-10 from MACHEREY-NAGEL, Germany: grain size 10 μm, pore diameter 30 nm, specific surface area 100 m^2 g^{-1}) is dried before use. The gel (10 g) is dispersed in dry toluene-pyridine solution (110 mL, toluene/pyridine, 10/1 by volume) followed by dropwise addition of trichlorovinylsilane (250 μL, 2.0 mmol) under nitrogen. After stirring the dispersion for 16 h at 50 °C, the silica gel is collected and washed successively with

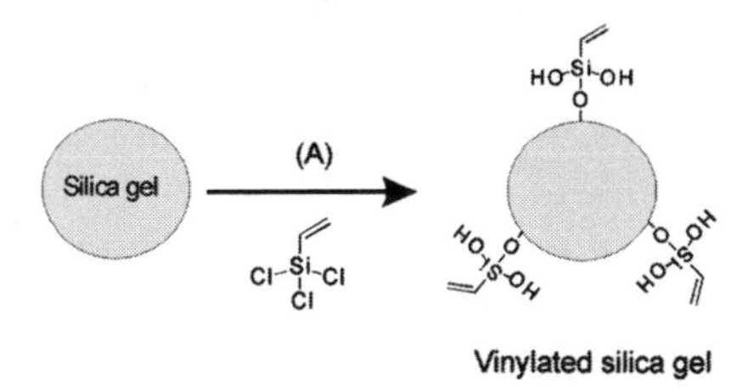

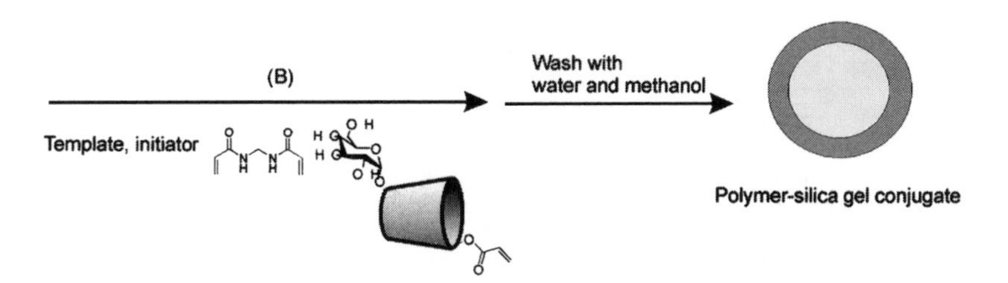

Fig. 8-2 Immobilization of the imprinted CyD polymer on the silica gel surface for the stationary phase of HPLC

chloroform, methanol, and water. Finally, the modified silica gel is dried under vacuum and used for the immobilization of the polymer. Since too much vinylation makes the gel immiscible with water, surface coverage should be maintained at about 10 %.

On this vinylated silica gel, CyD is imprinted (this technique is applicable to versatile water-soluble functional monomers). Acryloyl-CyD (90 mg, 67 μmol), MBAA (60 mg, 390 μmol), and the template molecule (30 μmol) are dissolved in 5 mM of Tris buffer solution (pH 8.0, 5 mL), and then vinylated silica gel (600 mg) is dispersed. The polymerization is started by adding potassium persulfate (7 μmol, 2 mg) and *N,N,N′,N′*-tetramethylethylenediamine (20 μmol, 3 μL) as an initiator system under nitrogen at 37 °C. After 1 h, the solid part is collected and washed with large amounts of water and methanol to remove the template and unreacted monomers. The polymer/silica-gel conjugates thus obtained are then packed into a stainless column and used as a stationary phase of HPLC. During the analysis, the column pressure is always kept normal.

Table 8.2 shows the retention behavior of two enantiomers of Phe-Phe. Although CyD itself is a chiral compound, the polymer/silica-gel conjugate prepared without template (Non-imp) discriminates between these enantiomers poorly. When L-Phe-L-Phe is used as template, however, L-Phe-L-Phe is retained more strongly than D-Phe-D-Phe. With the conjugate prepared in the presence of D-Phe-D-Phe, D-Phe-D-Phe is retained much more strongly than its enantiomer.

This new technique has two additional advantages. First, the amount of template required for the imprinting is far less (compare the experi-

Table 8-2 Retention times of Phe-Phe by the Phe-Phe-imprinted CyD polymers immobilized on the silica-gel support

Guest	*Retention time/min*		
	Non-Imp[a]	*L-Phe-L-Phe-Imp*	*D-Phe-D-Phe-Imp*
L-Phe-L-Phe	9.61	10.20	10.62
D-Phe-D-Phe	9.68	9.73	11.73

a Polymer/silica gel conjugate was prepared in the absence of template.

mental conditions for Examples 8.1 and 8.2). Second, the HPLC peaks are sufficiently sharp. Here, silica gel particles of uniform size are used, whereas, in conventional cases, imprinted polymers are mechanically crushed to the desired size.

8.2.2
Metal-Ligand Complexation for the Preorganization [5]

Polymer-bound metal ions can also be promising candidates for molecular recognition in water. If they are appropriately immobilized on the polymers by using molecular imprinting, the target guests can be selectively bound in water. For example, polymeric receptors for peptides are prepared by using Ni(II) ion (see Fig. 8.3). Two of the six coordination

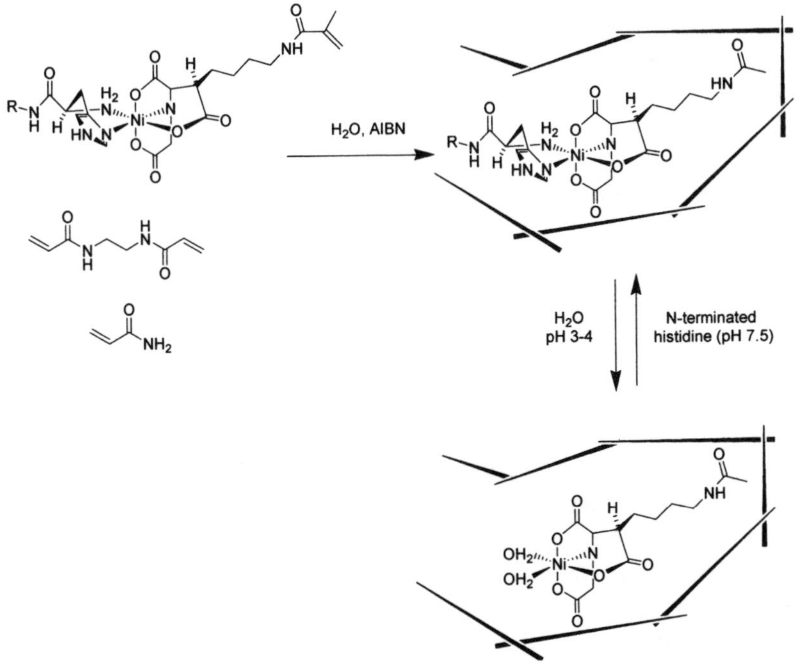

H_2O, AIBN

H_2O
pH 3-4

N-terminated
histidine (pH 7.5)

Fig. 8-3 Molecular imprinting of peptides in water by the use of metal complexation

sites of Ni(II) are used to bind the template at its *N*-terminal amine and the imidazole of *N*-terminal histidine. The acrylate residue is attached to the Ni(II) through the coordination of nitrilotriacetic acid. This complex is sufficiently stable in water at around pH 7, so that this preorganized complex can be directly copolymerized with acrylamide and *N,N'*-ethylenebisacrylamide. After the polymerization, the resultant polymer is treated with water at pH 3–4, and the template is removed. The residual Ni(II)-nitrilotriacetate complexes on the polymer serve as the specific binding sites for the template in water. The imprinting effect is clearly demonstrated by the large increase in capacity determined from the Scatchard plots. For instance, (His-Ala)-imprinted polymer has a significantly higher capacity for the template peptide over the other sequences such as His-Phe and His-Ala-Phe, indicating that the binding sites formed in the imprinted polymer are complementary to both size and shape. A non-histidine-containing peptide like Ala-Phe has almost no affinity for this polymer. This strategy is also promising for the recognition of peptides in water.

8.2.3
Molecular Imprinting at an Air-Water Interface

As described in the above sections, hydrogen bond formation in bulk water is difficult because of competition with the solvent molecules. But the air-water boundary is quite different from bulk water in physicochemical properties, and hydrogen bonding satisfactorily occurs there [6]. Recently, template polymerization of nucleoside analogs based on hydrogen bonding with nucleic acid has been achieved at this boundary by using the Langmuir-Blodgett technique [7, 8]. Natural oligonucleotides of predetermined sequence are dissolved in bulk water, and monolayers of photo-polymerizable (diacetylene group) amphiphile of nucleobase derivative are spread on its surface. At the air-water interface, amphiphiles are aggregated in an orderly fashion along the sequence of oligonucleotides in the bulk phase through hydrogen bond formation. These ordered aggregates are fixed by UV-irradiated photopolymerization.

The concept of »template polymerization« was proposed in 1970 to achieve precise control of the polymerization process. Since molecular recognition through multiple hydrogen bonding is strong and accurate enough for the arrangement of functional monomers, this is one of the promising strategies for the preparation of orderly polymers of functional residues.

8.3
Use of Two Kinds of Functional Monomers for Cooperative Recognition

In the general concept of molecular imprinting (Scheme 2.1), two or more functional monomers (circle- and square-shaped monomers) are placed complementarily to the template. However, most of the imprinted polymers hitherto reported are composed of only one kind of functional monomer, probably to avoid too complicated interactions among different monomers and template. In principle, the guest selectivity should be enhanced when plural monomers simultaneously bind the target molecule. One of the successful examples is the combination of a vinyl monomer of zinc porphyrin and methacrylic acid (MAA), which has been described in Example 3-6 in Chapter 3 [9]. The capacity factor is significantly enhanced when two different monomers are combined [compare PPM(imp) with PP(imp) or PM(imp) in Table 8.3]. Similar synergetic effects of these monomers are also evident when 9-ethyladenine (9-EA) is used as a template. By the simultaneous use of two monomers, the capacity factor for 9-EA is remarkable, while that for adenine (reference guest of 9-EA) is almost nil (Table 8.4) [10]. Although conventional imprinting using one kind of monomer promotes the selectivity, the imprinting effect is enormously enhanced by the second monomer.

Table 8-3 Retention properties of cinchonidine(CD)-imprinted and non-imprinted polymers obtained by using two kinds of functional monomers

Polymers	*Functional monomers*		*Template*	*Capacity factor* k [a]	
	Porphyrin	*MAA*		*CD*	*CN*
PPM(imp)	+	+	CD	12.86	6.37
PPM(non)	+	+	None	0.03	0.08
PP(imp)	+	–	CD	0.11	0.11
PP(non)	+	–	None	0.00	0.00
PM(imp)	–	+	CD	2.56	0.85
PM(non)	–	+	none	0.00	0.00

a $k = (t_R - t_0)/t_0$, where t_R is the retention time of a sample and t_0 is the time for a void marker. A mixture of CH_2Cl_2/MeOH/AcOH (91/6/3, v/v/v) is used as eluent.

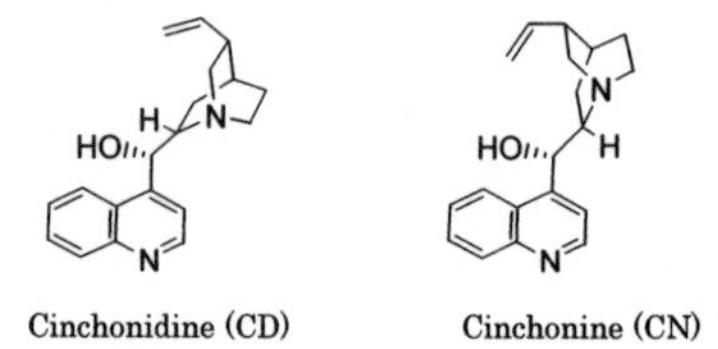

Cinchonidine (CD) Cinchonine (CN)

Table 8-4 Retention properties of the 9-ethyladenine(9-EA)-imprinted and non-imprinted polymers obtained by using two kinds of functional monomers

Polymers	*Functional monomers*		*Template*	*Capacity factor* k [a]	
	porphyrin	*MAA*		*9-EA*	*Adenine*
PPM(imp)	+	+	9-EA	28.9	0.0
PPM(non)	+	+	None	0.14	0.58
PP(imp)	+	–	9-EA	0.85	2.25
PP(non)	+	–	None	0.15	0.47
PM(imp)	–	+	9-EA	9.57	0.0
PM(non)	–	+	None	0.09	0.19

a $k = (t_R - t_0)/t_0$, where t_R is the retention time of a sample and t_0 is the time for a void marker. A mixture of CH_2Cl_2/MeOH/AcOH (97/2/1, v/v/v) is used as eluent.

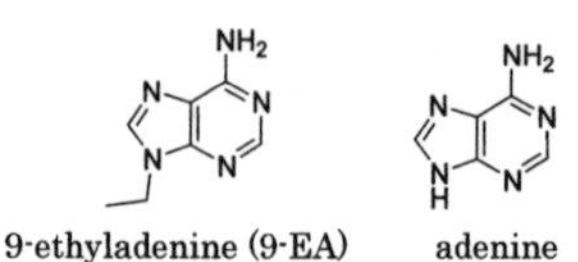

9-ethyladenine (9-EA) adenine

8.4
Inorganic Gel as the Matrix for Molecular Imprinting

Matrices used for molecular imprinting are not necessarily restricted to organic materials, and inorganic materials are also available [11, 12]. Modification of the surface of silica gel for the stationary phase of HPLC is known [13]. Recently, unique molecular imprinting by the use of inorganic materials showing various characteristic properties has been proposed.

8.4.1
Covalent Imprinting in Silica Gel Matrices

Inorganic gel oxides can be prepared from the corresponding alkoxides through sol-gel polycondensation. Silica gel is prepared from silicon tetraalkoxides. In this case, the alkoxides function as *crosslinkers* and make an oxide network through polycondensation. If some functional residues are retained complementarily to the specific target in the gel matrices, the gel oxides can be inorganic receptors. On the basis of this strategy, organic residues are introduced as binding sites by chemical modification of the silicon alkoxide, and can be retained in the silica gel after the polycondensation. For instance, the template is bound to the Si atom in silicon tetraalkoxides (e.g., tetraethoxysilane) by replacing one of the Si–O bonds with an Si–C bond. When silica gels are prepared by the hydrolysis of these Si–O bonds (and subsequent polycondensation of the resultant silanol), the Si–C bonds are maintained intact. Therefore, »covalent imprinting« is possible. In Fig. 8.4 [14], [3-(triethoxysilyl)propyl]-1,3,5-benzenetriyltris(methylene)carbamate (the template) is reacted with tetraethoxysilane (TEOS) in an acid-catalyzed sol-gel copolymerization (step a) to afford an organic-inorganic hybrid gel. After the reaction, the surface silanol groups are capped with trimethylsilyl groups (step b) by reacting with an equimolar mixture of chlorotrimethylsilane and 1,1,1,3,3,3-hexamethyldisilazane. Next, carbamate bonds are cleaved by reaction with trimethylsilyliodide, and the resulting trimethylsilylcarbamate is exposed to methanol and water to

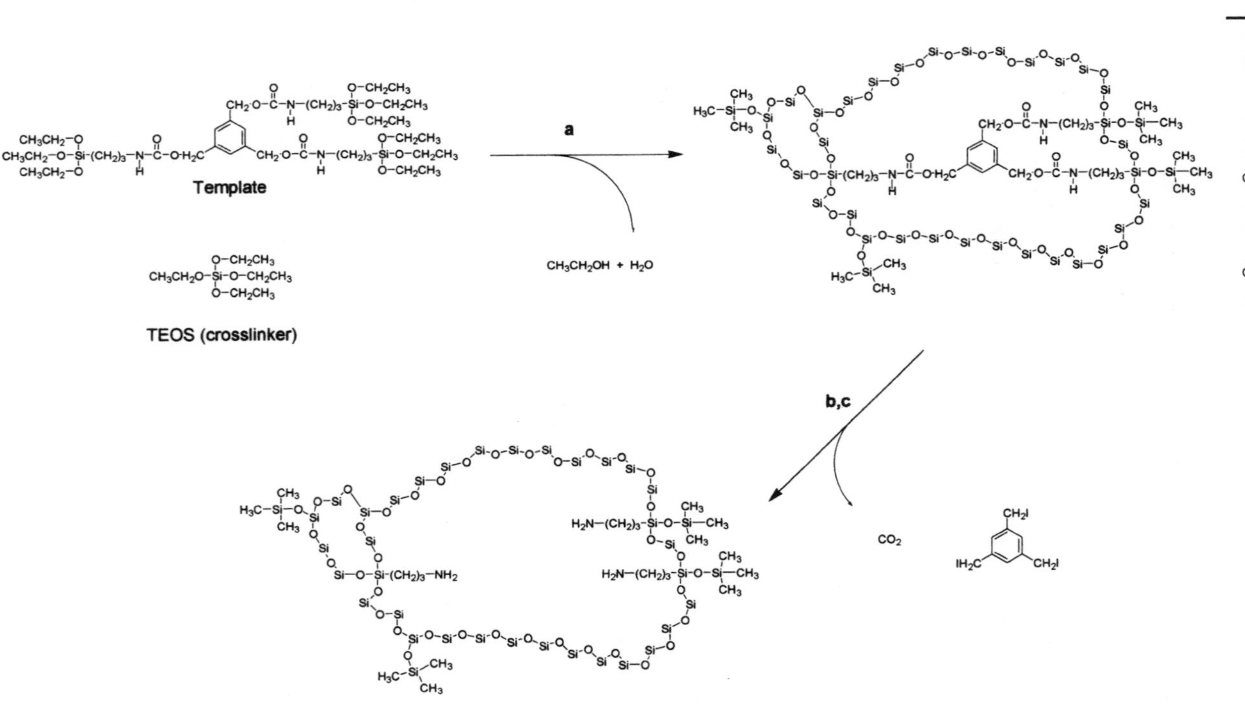

Fig. 8-4 Molecular imprinting in bulk silica gel: **a** sol-gel hydrolysis and polycondensation (pH 2.0), **b** capping of surface silanols with trimethylsilyliodide in acetonitrile, and **c** removal of the template with methanol/aqueous sodium bicarbonate

liberate the amines which act as guest-binding sites (step c). Oxide gels are also easily prepared from various metal alkoxides such as titanium (see Section 8.4.2), and various organic-inorganic hybrids are expected.

8.4.2
TiO_2 Ultrathin Film as a Matrix for Imprinting [15, 16]

As well as silicon alkoxides, titanium alkoxides are available as inorganic gel matrices. Since a liberated hydroxyl group (Ti-OH) in the gel can be a binding site, TiO_2 gel itself functions as an inorganic receptor. By applying the molecular imprinting technique to the preparation of TiO_2 gel, specific binding towards the target is possible.

Since the sol-gel process is applicable to the preparation of ultrathin film, molecularly imprinted inorganic films are synthesized from titanium alkoxide. Although guest binding cannot be analyzed by the conventional HPLC technique, quartz crystal microbalance (QCM) resonators satisfactorily detect the bound guest on the film as a frequency change in the nanogram range. Here, an ultrathin film composed of TiO_2 is formed directly on the all gold-coated QCM, and growth of the TiO_2 film as well as the guest binding is monitored. The actual procedure is shown below: a gold-coated QCM resonator modified with mercaptoethanol is immersed in $Ti(OBu)_4$ solution. Then the electrode is dipped in pure water to promote hydrolysis and condensation of chemisorbed alkoxide, and dried. Subsequently, it is immersed in aqueous template solution for the imprinting, followed by washing and drying. Then the electrode is again immersed in $Ti(OBu)_4$ solution. By repeating these procedures 10 times, the molecularly imprinted TiO_2 film schematically illustrated in Fig. 8.5 (total thickness of the film is about 14 nm as estimated from QCM data) is obtained. When dipeptide is used as a template, the TiO_2 film binds the template dipeptide strongly compared with other dipeptides.

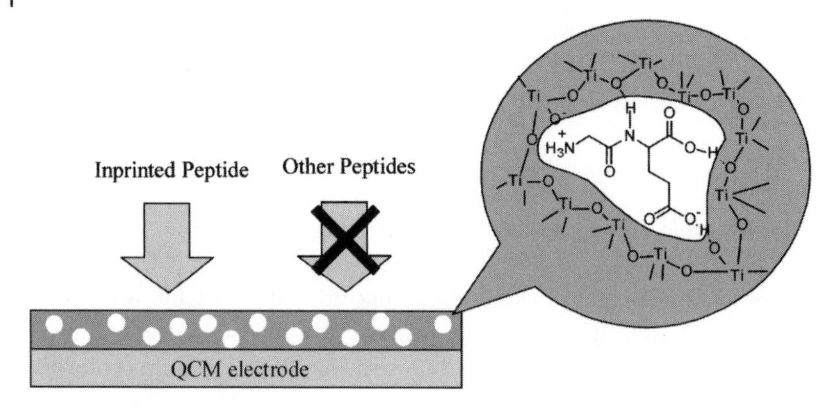

Fig. 8-5 Schematic illustration of a Gly-Glu-imprinted TiO_2-gel film

8.4.3
Helical Silica Gel by Molecular Imprinting [17]

Various artificial supramolecular structures of the nanometer to micron scale are easily created with *organic materials,* because we can design and synthesize the »building blocks« that comprise the supramolecules. But, in contrast with them, it is almost impossible to design and synthesize supramolecules from *inorganic materials* such as silica gel. These will not self-assemble into ordered structure and thus cannot have a specific shape. But, with the aid of organic materials, even inorganic materials can take on an ordered structure.

Recently, new organic gelators that can gelate various organic solvents have been developed. Some of these form a unique fibrous helical structure in certain solvents. Typical gelators with this property are shown in Fig. 8.6a. They can even gelate TEOS solution by forming a fibrous helical structure. As described in Section 8.4.1, TEOS is converted into silica gel through sol-gel polycondensation. Therefore, polycondensation of such a gelated TEOS solution affords silica gel with a helical hollow-fiber structure. In other words, a supramolecular structure of organic material is imprinted on silica-gel. By changing the organic gelators which form a specific supramolecular structure in gelated TEOS, we can imprint various supramolecular structures on inorganic silica gel. An actual procedure is shown below. Gelators 1 and 2 in Fig. 8.6a are dissolved in acetonitrile or ethanol, and then TEOS, water, and benzylamine as a polycondensation catalyst are added. At this stage, the solution is gelated. Then the solution is placed at room temperature under static conditions for 3-7 days, followed by drying by vacuum pump at room temperature. Subsequently, the gelator is removed by calcination at 200 °C for 2 h, 500 °C for 2 h under nitrogen atmosphere, and 500 °C for 4 h under aerobic conditions. SEM pictures of silica gel obtained thus are shown in Fig. 8.6b, demonstrating that helical fibrous silica-gel is formed.

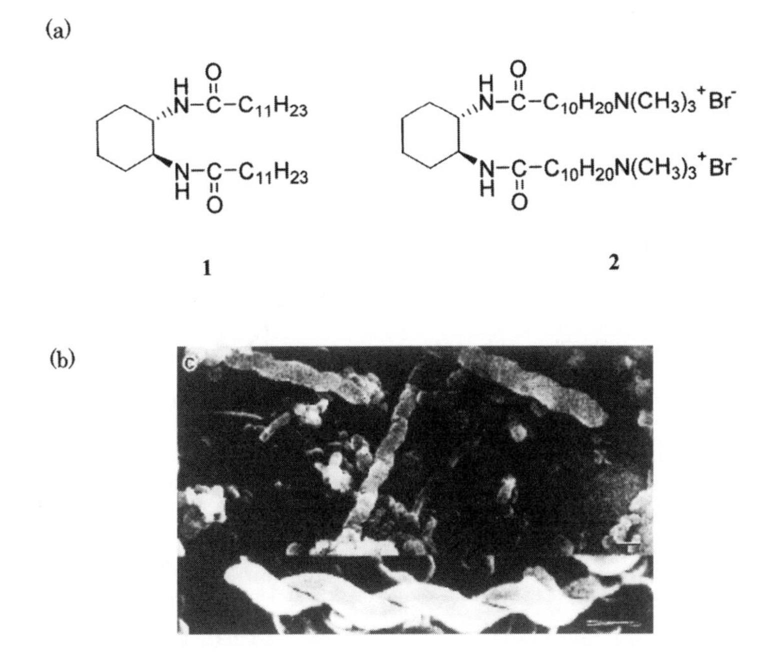

Fig. 8-6 Structure of typical organic gelators (a) and SEM picture of the silica (b) obtained from the mixed gelators (**1** + **2**) after calcination

8.5
Artificial Enzyme (Molecular Catalyst) by Molecular Imprinting

8.5.1
Conjugation of Catalytic Sites with Substrate-Binding Sites

Enzymes are basically composed of substrate-binding sites and catalytic sites which are close to each other and cooperatively accelerate the specific reactions. Their substrate specificity is interpreted in terms of the »lock and key« theory. Various artificial enzymes have been success-

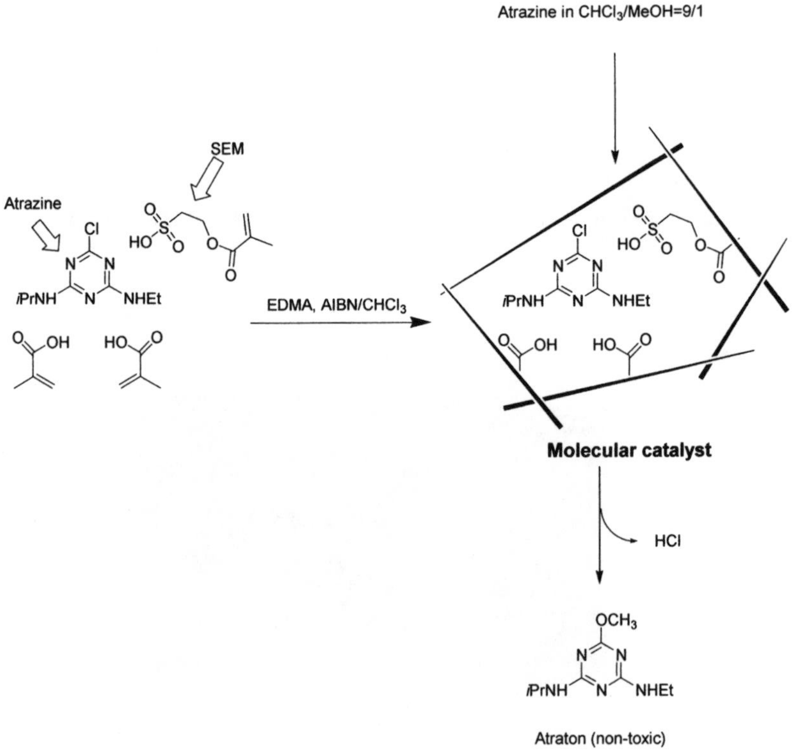

Fig. 8-7 Catalyst for the decomposition of atrazine synthesized by the molecular imprinting technique

fully synthesized by orderly placing of these two sites. The molecular imprinting method is very important here. For example, an artificial enzyme for the decomposition of atrazine, a herbicide giving considerable concern in current environmental science, is successfully synthesized by polymerizing MAA and 2-sulfoethyl methacrylate (SEM) in $CHCl_3$ in the presence of atrazine as template (see Fig. 8.7) [18]. The artificial enzyme obtained efficiently decomposes atrazine to atraton in the presence of methanol. Without the template, a much poorer catalyst is obtained. In this case, the catalytic site is the sulfoethyl group, which activates the methanol to allow the nucleophilic substitution on the triazine rings. Consistently, when MAA is polymerized in the absence of 2-sulfoethyl methacrylate, the polymer binds atrazine but shows no catalytic activity [19].

8.5.2 Catalytic Antibody Prepared by Using Transition-State Analog

In addition to the above conventional design of artificial enzymes, an entirely new concept has now appeared. As Pauling first pointed out, enzyme and antibody differ only in that the former binds the *transition state of a reaction* whereas the latter binds the *ground state* [20]. In other words, the antibody which selectively recognizes the transition state of a reaction should catalyze that reaction. These antibodies are called »catalytic antibodies«. This attractive idea has been elegantly substantiated. For example, ester hydrolysis proceeds through the transition state shown in Fig. 8.8a. Since this transition state itself is too unstable to use, its analog is synthesized by replacing the carbon at the center with phosphorus. Exactly as designed, the monoclonal antibody, prepared with this transition state analog as hapten (antigen), accelerates the hydrolysis of the corresponding carboxylic ester catalytically by a factor of 10^3–10^4 [21].

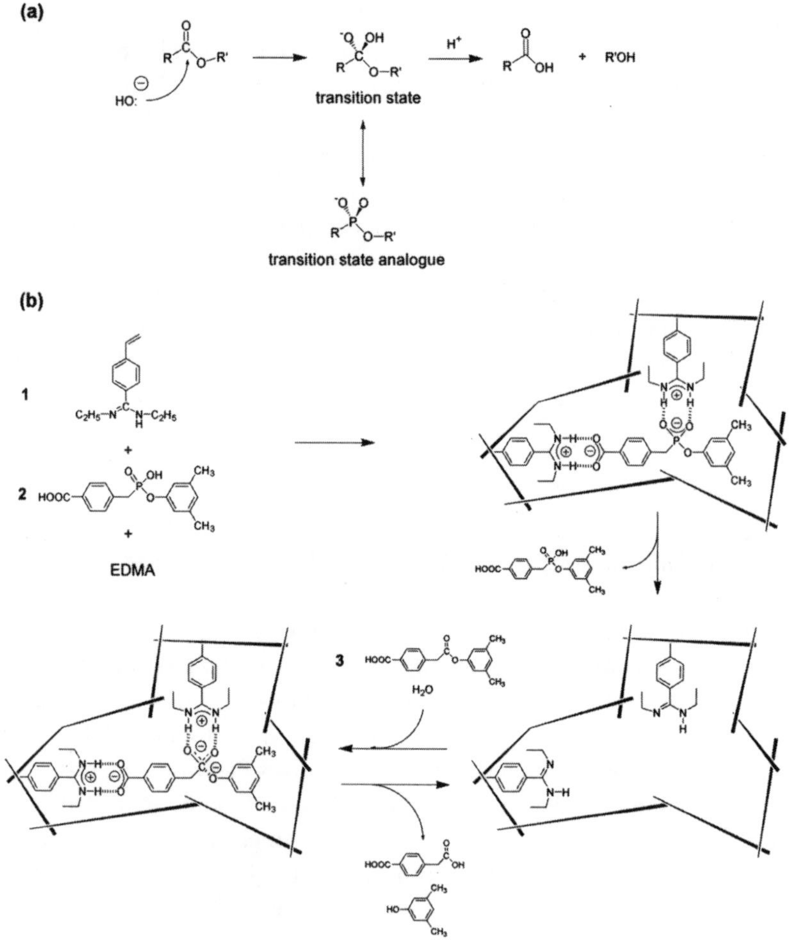

Figure 8-8 Mechanism of ester hydrolysis (**a**) and molecular imprinting of transition state analog (phosphonic acid (**2**) for mimicking catalytic antibody (**b**)

Example 8.3: Catalytic antibody as an artificial esterase

This new concept has been imported to molecular imprinting [22–25]. Ester hydrolysis is most conveniently used because (1) its reaction mechanism is well established, and (2) both substrate and transition state analogs are easy to obtain. In Fig. 8.8b, phosphonic acid (**2**) is used as a transition state analog of the hydrolysis of substrate **3** [26]. A vinyl monomer of amidine **1** is chosen as a functional monomer because it readily forms stable complexes with the carboxylic acid ester and the phosphonic acid monoester. The imprinted polymers are synthesized in THF from **1** (the monomer), **2** (the template), and ethylene glycol dimethacrylate (the cross-linker) by using AIBN as the radical initiator. After the polymerization, the template is removed by 0.1 N aqueous NaOH/acetonitrile (1/1) solution. The obtained polymer efficiently hydrolyzes **3** following the typical Michaelis-Menten kinetics, while the corresponding monomeric amidine solution hydrolyzes **3** about 100 times more slowly.

Example 8.4: Catalytic silica particles as an artificial enzyme *[27]*

A combination of organic and inorganic materials provides some interesting imprinted materials as demonstrated in Section 8.4. Similarly, a catalytic antibody can be prepared by organic-inorganic combination. Since accessibility of a substrate to the catalytic site crucially affects the reaction rate, the catalytic center should be located on the surface of silica gel. In other words, imprinting of transition state analog (TSA) is conducted on the surface. Here, a water-in-oil microemulsion offer a preferable environment for the surface imprinting, as schematically illustrated in Fig. 8.9. In the presence of surfactant, water is dispersed in oil, forming a stable emulsion, namely, reverse micelle. When a relatively hydrophilic TSA is connected to the head of hydrophobic long chain (surfactant TSA in Fig. 8.9) and is added with the surfactant, the »head« orients toward the water phase, keeping the long chain in the oil phase. Thus, TSA is located at the surface of the water phase. As described in Section 8.4, sol-gel polycondensation of silicon alkoxide proceeds through hydrolysis. By the addition of tetraethoxysilane (TEOS) and functionalized silane (including an organic functional group as a catalytic center), polycondensation occurs in the water emulsion phase to afford a silica particle whose surface is molecularly imprinted with the TSA (tetrahedral shape in Fig. 8.9). On the basis of this strategy, artificial protease is prepared by using an appropriate TSA for peptide hydrolysis.

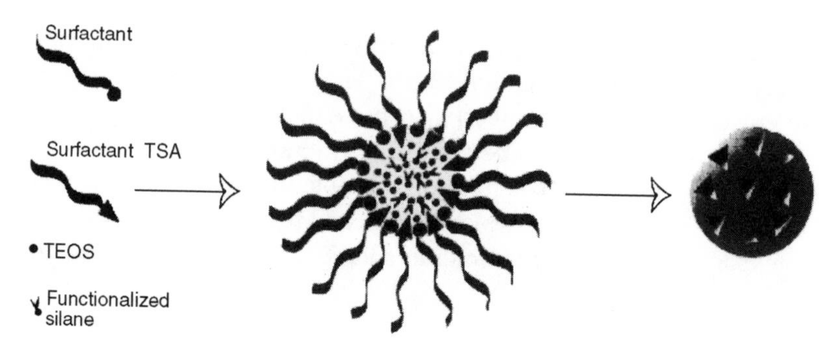

Fig. 8-9 Schematic illustration of the preparation of catalytic silica by surface imprinting in the presence of TSA as a template by use of the water-in-oil emulsion technique

References

1 Allender, C.J.; Brain, K.R.; Heard, C.M. »*Progress in Medicinal Chemistry*«, p. 235, Elsevier Science, Oxford (1999).

2 Bender, M.L.; Komiyama, M. »*Cyclodextrin Chemistry*«, Springer-Verlag, Berlin (1978).

3 Asanuma, H.; Akiyama, T.; Kajiya, K.; Hishiya, T.; Komiyama, M. *Anal. Chim. Acta.*, **2001**, *435*, 25–33.

4 Akiyama, T.; Hishiya, T.; Asanuma, H.; Komiyama, M. *J. Inclu. Phenom. Macrocyclic Chemistry*, **2001**, *41*, 149–153.

5 Hart, B.R.; Shea, K.J. *J. Am. Chem. Soc.*, **2001**, *123*, 2072–2073.

6 Kurihara, K.; Ohto, K.; Honda, Y.; Kunitake, T. *J. Am. Chem. Soc.* **1991**, *113*, 5077–5079.

7 Matsumoto, J.; Ijiro, K.; Shimomura, M. *Chem. Lett.*, **2000**, 1280–1281.

8 Ijiro, K.; Matsumoto, J.; Shimomura, M. *Studies in Surface Science and Catalysis*, **2001**, *132*, 481–484.

9 Takeuchi, T.; Mukawa, T.; Matsui, J.; Higashi, M.; Shimizu, K.D. *Anal. Chem.* **2001**, *73*, 3869–3874.

10 Matsui, J.; Higashi, M.; Takeuchi, T. *J. Am. Chem. Soc.* **2000**, *122*, 5218–5219.

11 Dickey, F.H. *Proc. Natl. Acad. Sci.* **1949**, *35*, 227–229.

12 Morihara, K.; Takiguchi, M.; Shimada, T. *Bull. Chem. Soc. Jpn.* **1994**, *67*, 1078–1084.

13 Wulff, G. *Angew. Chem. Int. Ed. Engl.* **1995**, *34*, 1812–1832.

14 Katz, A.; Davis, M.E. *Nature*, **2000**, *403*, 286–289.

15 Lee, S-W.; Ichinose, I.; Kunitake, T. *Langmuir*, **1998**, *14*, 2857–2863.

16 Ichinose, I.; Kikuchi, T.; Lee, S.-W.; Kunitake, T. *Chem. Lett.* **2002**, 104–105.

17 Jung, H.J.; Ono, Y.; Shinkai, S. *Chem. Eur. J.* **2000**, *6*, 4552–4557.

18 Takeuchi, T.; Fukuma, D.; Matsui, J.; Mukawa, T. *Chem. Lett.* **2001**, 530–531.

19 Matsui, J.; Miyoshi, Y.; Doblhoff-Dier, O.; Takeuchi, T. *Anal. Chem.* **1995**, *67*, 4404–4408.

20 Pauling, L. *Am. Sci.* **1948**, *36*, 51.

21 Lerner, R.A.; Benkovic, S.J.; Schultz, P.G. *Science*, **1991**, *252*, 659–667.

22 Matsui, J.; Nicholls, I.A.; Karube, I.; Mosbach, K. *J. Org. Chem.* **1996**, *61*, 5414–5417.

23 Morihara, K.; Kurihara, S.; Suzuki, J. *Bull. Chem. Soc. Jpn.* **1988**, *61*, 3991–3998.

24 Kawanami, Y.; Yunoki, T.; Nakamura, A.; Fujii, K.; Umano, K.; Yamauchi, H.; Masuda, K. *J. Mol. Catal. A* **1999**, *145*, 107–110.

25 Markowitz, M.A.; Kust, P.R.; Deng, G.; Schoen, P.E.; Dordick, J.S.; Clark, D.S.; Gaber, B.P. *Langmuir*, **2000**, *16*, 1759–1765.

26 Wulff G.; Gross T.; Schönfeld, R. *Angew. Chem. Int. Ed. Engl.* **1997**, *36*, 1962–1964.

27 Markowitz, M.A.; Kust, P.R.; Deng, G.; Schoen, P.E.; Dordick, J.S.; Clark, D.S.; Gaber, B.P. Langmuir, **2000**, *16*, 1759–1765.

Chapter 9
Conclusions and Prospects

For a long period of time, people have regarded solutions and gases as »simple aggregates of randomly moving molecules«. There has been no way to freeze the movements of these molecules and take a snapshot of their positions. The molecular imprinting method has for the first time made this dream come true. We can pick up any molecules in the system and place these target molecules on desired sites in the snapshot. Undoubtedly this method paves the way to the molecular devices (molecular memories, molecular machines, molecular computers, etc.) which are the keys to the advanced science of the 21st century. No other methods have ever provided important receptors so easily, promptly, and economically. Tailor-made receptors thus obtained protect our earth from environmental disruption. Other potential applications are too many to be counted.

Of course, the current molecular imprinting method has not yet been perfected with respect to its strictness in the »freezing«. The »shutter speed« of our camera must be further increased and the »shutter timing« must be more precisely controlled. We also have to know much more about the mechanism of imprinting. However, this area has been growing so rapidly in depth and breadth that these factors should be solved soon. No doubt, this versatile method will be still more widely used for a great many purposes in the near future. It is hoped that this book will help to achieve this goal.

Index